A TOUCH OF THE POET

A TOUCH OF THE POET

BY EUGENE O'NEILL

NELSON DOUBLEDAY, INC.
GARDEN CITY, NEW YORK

Manufactured in the United States of America

The original production of *A Touch of the Poet* opened at the Helen Hayes Theatre in New York City on October 2, 1958. Presented by The Producers Theatre (a Robert Whitehead production), it was staged by Harold Clurman and designed by Ben Edwards. The original cast, in order of appearance, was as follows:

MICKEY MALOY	TOM CLANCY
JAMIE CREGAN	CURT CONWAY
SARA MELODY	KIM STANLEY
NORA MELODY	HELEN HAYES
CORNELIUS MELODY	ERIC PORTMAN
DAN ROCHE	JOHN CALL
PADDY O'DOWD	ART SMITH
PATCH RILEY	FARRELL PELLY
DEBORAH (MRS. HENRY HARFORD)	BETTY FIELD
NICHOLAS GADSBY	LUDWIG VAN ROOTEN

A Touch of the Poet, presented by Elliott Martin by arrangement with the John F. Kennedy Center for the Performing Arts, opened on December 28, 1977, at the Helen Hayes Theatre in New York City. It was directed by José Quintero, setting and lighting were by Ben Edwards, costumes by Jane Greenwood. Production stage manager was Mitch Erickson and casting consultant was Marjorie Martin. The cast, in order of appearance, was as follows:

MICKEY MALOY	BARRY SNIDER
JAMIE CREGAN	MILO O'SHEA
SARA MELODY	KATHRYN WALKER
NORA MELODY	GERALDINE FITZGERALD
CORNELIUS MELODY	JASON ROBARDS
DAN ROCHE	WALTER FLANAGAN
PADDY O'DOWD	DERMOT MCNAMARA
PATCH RILEY	RICHARD HAMILTON
DEBORAH (MRS. HENRY HARFORD)	BETTY MILLER
NICHOLAS GADSBY	GEORGE EDE

A TOUCH OF THE POET

A Touch of the Poet takes place in the dining room of Melody's Tavern, in a village a few miles from Boston.

Act One	Morning of July 27, 1828.
Act Two	Later that morning.
Act Three	That evening.
Act Four	That night.

Act One

SCENE *The dining room of Melody's Tavern, in a village a few miles from Boston. The tavern is over a hundred years old. It had once been prosperous, a breakfast stop for the stagecoach, but the stage line had been discontinued and for some years now the tavern has fallen upon neglected days.*

The dining room and barroom were once a single spacious room, low-ceilinged, with heavy oak beams and paneled walls—the taproom of the tavern in its prosperous days, now divided into two rooms by a flimsy partition, the barroom being off left. The partition is painted to imitate the old paneled walls but this only makes it more of an eyesore.

At left front, two steps lead up to a closed door opening on a flight of stairs to the floor above. Farther back is the door to the bar. Between these doors hangs a large mirror. Beyond the bar door a small cabinet is fastened to the wall. At rear are four windows. Between the middle two is the street door. At right front is another door, open, giving on a hallway and the main stairway to the second floor, and leading to the kitchen. Farther front at right, there is a high schoolmaster's desk with a stool.

In the foreground are two tables. One, with four chairs, at left center; a larger one, seating six, at right center. At left and right, rear, are two more tables, identical with the ones at right center. All these tables are set with white tablecloths, etc., except the small ones in the foreground at left.

It is around nine in the morning of July 27, 1828. Sunlight shines in through the windows at rear.

Mickey Maloy sits at the table at left front, facing right. He is glancing through a newspaper. Maloy is twenty-six, with a sturdy physique and an amiable, cunning face, his mouth usually set in a half-leering grin.

Jamie Cregan peers around the half-open door to the bar. Seeing Maloy, he comes in. As obviously Irish as Maloy, he is middle-aged, tall, with a lantern-jawed face. There is a scar of a saber cut over one cheekbone. He is dressed neatly but in old, worn clothes. His eyes are bloodshot, his manner sickly, but he grins as he greets Maloy sardonically.

CREGAN

God bless all here—even the barkeep.

MALOY

With an answering grin.

Top o' the mornin'.

CREGAN

Top o' me head.

He puts his hand to his head and groans.

Be the saints, there's a blacksmith at work on it!

MALOY

Small wonder. You'd the divil's own load when you left at two this mornin'.

CREGAN

I must have. I don't remember leaving.

He sits at right of table.

Faix, you're takin' it aisy.

MALOY

There's no trade this time o' day.

CREGAN

It was a great temptation, when I saw no one in the bar, to make off with a bottle. A hair av the dog is what I need, but I've divil a penny in my pantaloons.

MALOY

Have one on the house.

He goes to the cupboard and takes out a decanter of whiskey and a glass.

CREGAN

Thank you kindly. Sure, the good Samaritan was a crool haythen beside you.

MALOY

Putting the decanter and glass before him.

It's the same you was drinking last night—his private dew. He keeps it here for emergencies when he don't want to go in the bar.

CREGAN

Pours out a big drink.

Lave it to Con never to be caught dry.

Raising his glass.

Your health and inclinations—if they're virtuous!

He drinks and sighs with relief.

God bless you, Whiskey, it's you can rouse the dead! Con hasn't been down yet for his morning's morning?

MALOY

No. He won't be till later.

CREGAN

It's like a miracle, me meeting him again. I came to these parts looking for work. It's only by accident I heard talk of a Con

Melody and come here to see was it him. Until last night, I'd not seen hide nor hair of him since the war with the French in Spain—after the battle of Salamanca in '12. I was a corporal in the Seventh Dragoons and he was major.

Proudly

I got this cut from a saber at Talavera, bad luck to it! —serving under him. He was a captain then.

MALOY

So you told me last night.

CREGAN

With a quick glance at him.

Did I now? I must have said more than my prayers, with the lashings of whiskey in me.

MALOY

With a grin.

More than your prayers is the truth.

Cregan glances at him uneasily. Maloy pushes the decanter toward him.

Take another taste.

CREGAN

I don't like sponging. Sure, my credit ought to be good in this shebeen! Ain't I his cousin?

MALOY

You're forgettin' what himself told you last night as he went up to bed. You could have all the whiskey you could pour down you, but not a penny's worth of credit. This house, he axed you to remember, only gives credit to gentlemen.

CREGAN

Divil mend him!

MALOY

With a chuckle.

You kept thinking about his insults after he'd gone out, getting madder and madder.

CREGAN

God pity him, that's like him. He hasn't changed much.

He pours out a drink and gulps it down—with a cautious look at Maloy.

If I was mad at Con, and me blind drunk, I must have told you a power of lies.

MALOY

Winks slyly.

Maybe they wasn't lies.

CREGAN

If I said any wrong of Con Melody—

MALOY

Arrah, are you afraid I'll gab what you said to him? I won't, you can take my oath.

CREGAN

His face clearing.

Tell me what I said and I'll tell you if it was lies.

MALOY

You said his father wasn't of the quality of Galway like he makes out, but a thievin' shebeen keeper who got rich by moneylendin' and squeezin' tenants and every manner of trick. And when he'd enough he married, and bought an estate with a pack of hounds and set up as one of the gentry. He'd hardly got settled when his wife died givin' birth to Con.

CREGAN

There's no lie there.

MALOY

You said none of the gentry would speak to auld Melody, but he had a tough hide and didn't heed them. He made up his mind he'd bring Con up a true gentleman, so he packed him off to Dublin to school, and after that to the College with sloos of money to prove himself the equal of any gentleman's son. But Con found, while there was plenty to drink on him and borrow money, there was few didn't sneer behind his back at his pretensions.

CREGAN

That's the truth, too. But Con wiped the sneer off their mugs when he called one av thim out and put a bullet in his hip. That was his first duel. It gave his pride the taste for revenge and after that he was always lookin' for an excuse to challenge someone.

MALOY

He's done a power av boastin' about his duels, but I thought he was lyin'.

CREGAN

There's no lie in it. It was that brought disgrace on him in the end, right after he'd been promoted to major. He got caught by a Spanish noble making love to his wife, just after the battle of Salamanca, and there was a duel and Con killed him. The scandal was hushed up but Con had to resign from the army. If it wasn't for his fine record for bravery in battle, they'd have court-martialed him.

Then guiltily.

But I'm sayin' more than my prayers again.

MALOY

It's no news about his women. You'd think, to hear him when he's drunk, there wasn't one could resist him in Portugal and Spain.

CREGAN

If you'd seen him then, you wouldn't wonder. He was as strong as an ox, and on a thoroughbred horse, in his uniform, there wasn't a handsomer man in the army. And he had the chance he wanted in Portugal and Spain where a British officer was welcome in the gentry's houses. At home, the only women he'd known was whores.

He adds hastily.

Except Nora, I mean.

Lowering his voice.

Tell me, has he done any rampagin' wid women here?

MALOY

He hasn't. The damned Yankee gentry won't let him come near them, and he considers the few Irish around here to be scum beneath his notice. But once in a while there'll be some Yankee stops overnight wid his wife or daughter and then you'd laugh to see Con, if he thinks she's gentry, sidlin' up to her, playin' the great gentleman and makin' compliments, and then boasting afterward he could have them in bed if he'd had a chance at it, for all their modern Yankee airs.

CREGAN

And maybe he could. If you'd known him in the auld days, you'd nivir doubt any boast he makes about fightin' and women, and gamblin' or any kind av craziness. There nivir was a madder divil.

MALOY

Lowering his voice.

Speakin' av Nora, you nivir mentioned her last night, but I know all about it without you telling me. I used to have my room here, and there's nights he's madder drunk than most when he throws it in her face he had to marry her because— Mind you, I'm not saying anything against poor Nora. A sweeter woman never lived. And I know you know all about it.

CREGAN

Reluctantly.

I do. Wasn't I raised on his estate?

MALOY

He tells her it was the priests tricked him into marrying her. He hates priests.

CREGAN

He's a liar, then. He may like to blame it on them but it's little Con Melody cared what they said. Nothing ever made him do anything, except himself. He married her because he'd fallen in love with her, but he was ashamed of her in his pride at the same time because her folks were only ignorant peasants on his estate, as poor as poor. Nora was as pretty a girl as you'd find in a year's travel, and he'd come to be bitter lonely, with no woman's company but the whores was helpin' him ruin the estate.

He shrugs his shoulders.

Well, anyways, he married her and then went off to the war, and left her alone in the castle to have her child, and nivir saw her again till he was sent home from Spain. Then he raised what money he still was able, and took her and Sara here to America where no one would know him.

MALOY

Thinking this over for a moment.

It's hard for me to believe he ever loved her. I've seen the way he treats her now. Well, thank you for telling me, and I take my oath I'll nivir breathe a word of it—for Nora's sake, not his.

CREGAN

Grimly.

You'd better kape quiet for fear of him, too. If he's one-half the man he was, he could bate the lights out of the two av us.

MALOY

He's strong as a bull still for all the whiskey he's drunk.

He pushes the bottle toward Cregan.

Have another taste.

Cregan pours out a drink.

Drink hearty.

CREGAN

Long life.

He drinks. Maloy puts the decanter and glass back on the cupboard. A girl's voice is heard from the hall at right. Cregan jumps up—hastily.

That's Sara, isn't it? I'll get out. She'll likely blame me for Con getting so drunk last night. I'll be back after Con is down.

He goes out. Maloy starts to go in the bar, as if he too wanted to avoid Sara. Then he sits down defiantly.

MALOY

Be damned if I'll run from her.

He takes up the paper as Sara Melody comes in from the hall at right.

Sara is twenty, an exceedingly pretty girl with a mass of black hair, fair skin with rosy cheeks, and beautiful, deep-blue eyes. There is a curious blending in her of what are commonly considered aristocratic and peasant characteristics. She has a fine forehead. Her nose is thin and straight. She has small ears set close to her well-shaped head, and a slender neck. Her mouth, on the other hand, has a touch of coarseness and sensuality and her jaw is too heavy. Her figure is strong and graceful, with full, firm breasts and hips, and a slender waist. But she has large feet and broad, ugly hands with stubby fingers. Her voice is soft and musical, but her speech has at times a self-conscious, stilted quality about it, due to her restraining a tendency to lapse into brogue. Her everyday working dress is of cheap material, but she wears it in a way that gives a pleasing effect of beauty unadorned.

SARA

With a glance at Maloy, sarcastically.

I'm sorry to interrupt you when you're so busy, but have you your bar book ready for me to look over?

MALOY

Surlily.

I have. I put it on your desk.

SARA

Thank you.

She turns her back on him, sits at the desk, takes a small account book from it, and begins checking figures.

MALOY

Watches her over his paper.

If it's profits you're looking for, you won't find them—not with all the drinks himself's been treating to.

She ignores this. He becomes resentful.

You've got your airs of a grand lady this morning, I see. There's no talkin' to you since you've been playin' nurse to the young Yankee upstairs.

She makes herself ignore this, too.

Well, you've had your cap set for him ever since he came to live by the lake, and now's your chance, when he's here sick and too weak to defend himself.

SARA

Turns on him—with quiet anger.

I warn you to mind your own business, Mickey, or I'll tell my father of your impudence. He'll teach you to keep your place, and God help you.

MALOY

Doesn't believe this threat but is frightened by the possibility.

Arrah, don't try to scare me. I know you'd never carry tales to him.

Placatingly.

Can't you take a bit of teasing, Sara?

SARA

Turns back to her figuring.

Leave Simon out of your teasing.

MALOY

Oho, he's Simon to you now, is he? Well, well.

He gives her a cunning glance.

Maybe, if you'd come down from your high horse, I could tell you some news.

SARA

You're worse than an old woman for gossip. I don't want to hear it.

MALOY

When you was upstairs at the back taking him his breakfast, there was a grand carriage with a nigger coachman stopped at the corner and a Yankee lady got out and came in here. I was sweeping and Nora was scrubbing the kitchen.

Sara has turned to him, all attention now.

She asked me what road would take her near the lake—

SARA

Starts.

Ah.

MALOY

So I told her, but she didn't go. She kept looking around, and said she'd like a cup of tea, and where was the waitress. I knew she must be connected someway with Harford or why would she want to go to the lake, where no one's ever lived but him. She didn't want tea at all, but only an excuse to stay.

SARA

Resentfully.

So she asked for the waitress, did she? I hope you told her I'm the owner's daughter, too.

MALOY

I did. I don't like Yankee airs any more than you. I was short with her. I said you was out for a walk, and the tavern wasn't open yet, anyway. So she went out and drove off.

SARA

Worriedly now.

I hope you didn't insult her with your bad manners. What did she look like, Mickey?

MALOY

Pretty, if you like that kind. A pale, delicate wisp of a thing with big eyes.

SARA

That fits what he's said of his mother. How old was she?

MALOY

It's hard to tell, but she's too young for his mother, I'd swear. Around thirty, I'd say. Maybe it's his sister.

SARA

He hasn't a sister.

MALOY

Grinning.

Then maybe she's an old sweetheart looking for you to scratch your eyes out.

SARA

He's never had a sweetheart.

MALOY

Mockingly.

Is that what he tells you, and you believe him? Faix, you must be in love!

SARA

Angrily.

Will you mind your own business? I'm not such a fool!

Worried again.

Maybe you ought to have told her he's here sick to save her the drive in the hot sun and the walk through the woods for nothing.

MALOY

Why would I tell her, when she never mentioned him?

SARA

Yes, it's her own fault. But– Well, there's no use thinking of it now–or bothering my head about her, anyway, whoever she was.

She begins checking figures. Her mother appears in the doorway at right.

Nora Melody is forty, but years of overwork and worry have made her look much older. She must have been as pretty as a girl as Sara is now. She still has the beautiful eyes her daughter has inherited. But she has become too worn out to take care of her appearance. Her black hair, streaked with gray, straggles in untidy wisps about her face. Her body is dumpy, with sagging breasts, and her old clothes are like a bag covering it, tied around the middle. Her red hands are knotted by rheumatism. Cracked working shoes, run down at the heel, are on her bare feet. Yet in spite of her slovenly appearance there is a spirit which shines through and makes her lovable, a simple sweetness and charm, something gentle and sad and, somehow, dauntless.

MALOY

Jumps to his feet, his face lighting up with affection.

God bless you, Nora, you're the one I was waitin' to see. Will you keep an eye on the bar while I run to the store for a bit av 'baccy?

SARA

Sharply.

Don't do it, Mother.

NORA

Smiles—her voice is soft, with a rich brogue.

Why wouldn't I? "Don't do it, Mother."

MALOY

Thank you, Nora.

He goes to the door at rear and opens it, burning for a parting shot at Sara.

And the back o' my hand to you, your Ladyship!

He goes out, closing the door.

SARA

You shouldn't encourage his laziness. He's always looking for excuses to shirk.

NORA

Ah, nivir mind, he's a good lad.

She lowers herself painfully on the nearest chair at the rear of the table at center front.

Bad cess to the rheumatism. It has me destroyed this mornin'.

SARA

Still checking figures in the book—gives her mother an impatient but at the same time worried glance.

Her habitual manner toward her is one of mingled love and pity and exasperation.

I've told you a hundred times to see the doctor.

NORA

We've no money for doctors. They're bad luck, anyway. They bring death with them.

A pause. Nora sighs.

Your father will be down soon. I've some fine fresh eggs for his breakfast.

SARA

Her face becomes hard and bitter.

He won't want them.

NORA

Defensively.

You mean he'd a drop too much taken last night? Well, small blame to him, he hasn't seen Jamie since—

SARA

Last night? What night hasn't he?

NORA

A pause—worriedly.

Neilan sent round a note to me about his bill. He says we'll have to settle by the end of the week or we'll get no more groceries.

With a sigh.

I can't blame him. How we'll manage, I dunno. There's the intrist on the mortgage due the first. But that I've saved, God be thanked.

SARA

Exasperatedly.

If you'd only let me take charge of the money.

NORA

With a flare of spirit.

I won't. It'd mean you and himself would be at each other's throats from dawn to dark. It's bad enough between you as it is.

SARA

Why didn't you pay Neilan the end of last week? You told me you had the money put aside.

NORA

So I did. But Dickinson was tormentin' your father with his feed bill for the mare.

SARA

Angrily.

I might have known! The mare comes first, if she takes the bread out of our mouths! The grand gentleman must have his thoroughbred to ride out in state!

NORA

Defensively.

Where's the harm? She's his greatest pride. He'd be heart-broken if he had to sell her.

SARA

Oh yes, I know well he cares more for a horse than for us!

NORA

Don't be saying that. He has great love for you, even if you do be provokin' him all the time.

SARA

Great love for me! Arrah, God pity you, Mother!

NORA

Sharply.

Don't put on the brogue, now. You know how he hates to hear you. And I do, too. There's no excuse not to cure yourself. Didn't he send you to school so you could talk like a gentleman's daughter?

SARA

Resentfully, but more careful of her speech.

If he did, I wasn't there long.

NORA

It was you insisted on leavin'.

SARA

Because if he hadn't the pride or love for you not to live on your slaving your heart out, I had that pride and love!

NORA

Tenderly.

I know, Acushla. I know.

SARA

With bitter scorn.

We can't afford a waitress, but he can afford to keep a thoroughbred mare to prance around on and show himself off! And he can afford a barkeep when, if he had any decency, he'd do his part and tend the bar himself.

NORA

Indignantly.

Him, a gentleman, tend bar!

SARA

A gentleman! Och, Mother, it's all right for the two of us, out of our own pride, to pretend to the world we believe that lie, but it's crazy for you to pretend to me.

NORA

Stubbornly.

It's no lie. He *is* a gentleman. Wasn't he born rich in a castle on a grand estate and educated in college, and wasn't he an officer in the Duke of Wellington's army—

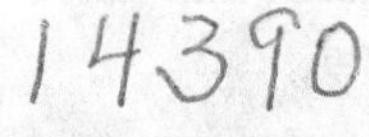

SARA

All right, Mother. You can humor his craziness, but he'll never make me pretend to him I don't know the truth.

NORA

Don't talk as if you hated him. You ought to be shamed—

SARA

I do hate him for the way he treats you. I heard him again last night, raking up the past, and blaming his ruin on his having to marry you.

NORA

Protests miserably.

It was the drink talkin', not him.

SARA

Exasperated.

It's you ought to be ashamed, for not having more pride! You bear all his insults as meek as a lamb! You keep on slaving for him when it's that has made you old before your time!

Angrily.

You can't much longer, I tell you! He's getting worse. You'll have to leave him.

NORA

Aroused.

I'll never! Howld your prate!

SARA

You'd leave him today, if you had any pride!

NORA

I've pride in my love for him! I've loved him since the day I set eyes on him, and I'll love him till the day I die!

With a strange superior scorn.

It's little you know of love, and you never will, for there's the same divil of pride in you that's in him, and it'll kape you from ivir givin' all of yourself, and that's what love is.

SARA

I could give all of myself if I wanted to, but—

NORA

If! Wanted to! Faix, it proves how little of love you know when you prate about if's and want-to's. It's when you don't give a thought for all the if's and want-to's in the world! It's when, if all the fires of hell was between you, you'd walk in them gladly to be with him, and sing with joy at your own burnin', if only his kiss was on your mouth! That's love, and I'm proud I've known the great sorrow and joy of it!

SARA

Cannot help being impressed—looks at her mother with wondering respect.

You're a strange woman, Mother.

She kisses her impulsively.

And a grand woman!

Defiant again, with an arrogant toss of her head.

I'll love—but I'll love where it'll gain me freedom and not put me in slavery for life.

NORA

There's no slavery in it when you love!

Suddenly her exultant expression crumbles and she breaks down.

For the love of God, don't take the pride of my love from me, Sara, for without it what am I at all but an ugly, fat woman gettin' old and sick!

SARA

Puts her arm around her—soothingly.

Hush, Mother. Don't mind me.

Briskly, to distract her mother's mind.

I've got to finish the bar book. Mickey can't put two and two together without making five.

She goes to the desk and begins checking figures again.

NORA

Dries her eyes—after a pause she sighs worriedly.

I'm worried about your father. Father Flynn stopped me on the road yesterday and tould me I'd better warn him not to sneer at the Irish around here and call thim scum, or he'll get in trouble. Most of thim is in a rage at him because he's come out against Jackson and the Democrats and says he'll vote with the Yankees for Quincy Adams.

SARA

Contemptuously.

Faith, they can't see a joke, then, for it's a great joke to hear him shout against mob rule, like one of the Yankee gentry, when you know what he came from. And after the way the Yanks swindled him when he came here, getting him to buy this inn by telling him a new coach line was going to stop here.

She laughs with bitter scorn.

Oh, he's the easiest fool ever came to America! It's that I hold against him as much as anything, that when he came here the chance was before him to make himself all his lies pretended to be. He had education above most Yanks, and he had money enough to start him, and this is a country where you can rise as high as you like, and no one but the fools who envy you care what you rose from, once you've the money and the power goes with it.

Passionately.

Oh, if I was a man with the chance he had, there wouldn't be a dream I'd not make come true!

She looks at her mother, who is staring at the floor dejectedly and hasn't been listening. She is exasperated for a second—then she smiles pityingly.

You're a fine one to talk to, Mother. Wake up. What's worrying you now?

NORA

Father Flynn tould me again I'd be damned in hell for lettin' your father make a haythen of me and bring you up a haythen, too.

SARA

With an arrogant toss of her head.

Let Father Flynn mind his own business, and not frighten you with fairy tales about hell.

NORA

It's true, just the same.

SARA

True, me foot! You ought to tell the good Father we aren't the ignorant shanty scum he's used to dealing with.

She changes the subject abruptly—closing Mickey's bar book.

There. That's done.

She puts the book in the desk.

I'll take a walk to the store and have a talk with Neilan. Maybe I can blarney him to let the bill go another month.

NORA

Gratefully.

Oh, you can. Sure, you can charm a bird out of a tree when you want to. But I don't like you beggin' to a Yankee. It's all right for me but I know how you hate it.

SARA

Puts her arms around her mother—tenderly.

I don't mind at all, if I can save you a bit of the worry that's killing you.

She kisses her.

I'll change to my Sunday dress so I can make a good impression.

NORA

With a teasing smile.

I'm thinkin' it isn't on Neilan alone you want to make an impression. You've changed to your Sunday best a lot lately.

SARA

Coquettishly.

Aren't you the sly one! Well, maybe you're right.

NORA

How was he went you took him his breakfast?

SARA

Hungry, and that's a good sign. He had no fever last night. Oh, he's on the road to recovery now, and it won't be long before he'll be back in his cabin by the lake.

NORA

I'll never get it clear in my head what he's been doing there the past year, living like a tramp or a tinker, and him a rich gentleman's son.

SARA

With a tender smile.

Oh, he isn't like his kind, or like anyone else at all. He's a born dreamer with a raft of great dreams, and he's very serious about them. I've told you before he wanted to get away from his father's business, where he worked for a year after he graduated from Harvard College, because he didn't like being in trade, even if it is a great company that trades with the whole world in its own ships.

NORA

Approvingly.

That's the way a true gentleman would feel—

SARA

He wanted to prove his independence by living alone in the wilds, and build his own cabin, and do all the work, and support himself simply, and feel one with Nature, and think great thoughts about what life means, and write a book about how the world can be changed so people won't be greedy to own money and land and get the best of each other but will be content with little and live in peace and freedom together, and it will be like heaven on earth.

She laughs fondly—and a bit derisively.

I can't remember all of it. It seems crazy to me, when I think of what people are like. He hasn't written any of it yet, anyway—only the notes for it.

She smiles coquettishly.

All he's written the last few months are love poems.

NORA

That's since you began to take long walks by the lake.

She smiles.

It's you are the sly one.

SARA

Laughing.

Well, why shouldn't I take walks on our own property?

Her tone changes to a sneer.

The land our great gentleman was swindled into buying when he came here with grand ideas of owning an American estate! —a bit of farm land no one would work any more, and the rest all wilderness! You couldn't give it away.

NORA

Soothingly.

Hush, now.

Changing the subject.

Well, it's easy to tell young Master Harford has a touch av the poet in him—

She adds before she thinks.

the same as your father.

SARA

Scornfully.

God help you, Mother! Do you think Father's a poet because he shows off reciting Lord Byron?

NORA

With an uneasy glance at the door at left front.

Whist, now. Himself will be down any moment.

Changing the subject.

I can see the Harford lad is falling in love with you.

SARA

Her face lights up triumphantly.

Falling? He's fallen head over heels. He's so timid, he hasn't told me yet, but I'll get him to soon.

NORA

I know you're in love with him.

SARA

Simply.

I am, Mother.

She adds quickly.

But not too much. I'll not let love make me any man's slave. I want to love him just enough so I can marry him without cheating him, or myself.

Determinedly.

For I'm going to marry him, Mother. It's my chance to rise in the world and nothing will keep me from it.

NORA

Admiringly.

Musha, but you've boastful talk! What about his fine Yankee family? His father'll likely cut him off widout a penny if he marries a girl who's poor and Irish.

SARA

He may at first, but when I've proved what a good wife I'll be— He can't keep Simon from marrying me. I know that.

Simon doesn't care what his father thinks. It's only his mother I'm afraid of. I can tell she's had great influence over him. She must be a queer creature, from all he's told me. She's very strange in her ways. She never goes out at all but stays home in their mansion, reading books, or in her garden.

She pauses.

Did you notice a carriage stop here this morning, Mother?

NORA

Preoccupied–uneasily.

Don't count your chickens before they're hatched. Young Harford seems a dacent lad. But maybe it's not marriage he's after.

SARA

Angrily.

I won't have you wronging him, Mother. He has no thought–

Bitterly.

I suppose you're bound to suspect–

She bites her words back, ashamed.

Forgive me, Mother. But it's wrong of you to think badly of Simon.

She smiles.

You don't know him. Faith, if it came to seducing, it'd be me that'd have to do it. He's that respectful you'd think I was a holy image. It's only in his poems, and in the diary he keeps– I had a peek in it one day I went to tidy up the cabin for him. He's terribly ashamed of his sinful inclinations and the insult they are to my purity.

She laughs tenderly.

NORA

Smiling, but a bit shocked.

Don't talk so bould. I don't know if it's right, you to be in his room so much, even if he is sick. There's a power av talk about the two av you already.

SARA

Let there be, for all I care! Or all Simon cares either. When it comes to not letting others rule him, he's got a will of his own behind his gentleness. Just as behind his poetry and dreams I feel he has it in him to do anything he wants. So even if his father cuts him off, with me to help him we'll get on in the world. For I'm no fool, either.

NORA

Glory be to God, you have the fine opinion av yourself!

SARA

Laughing.

Haven't I, though!

Then bitterly.

I've had need to have, to hold my head up, slaving as a waitress and chambermaid so my father can get drunk every night like a gentleman!

The door at left front is slowly opened and Cornelius Melody appears in the doorway above the two steps. He and Sara stare at each other. She stiffens into hostility and her mouth sets in scorn. For a second his eyes waver and he looks guilty. Then his face becomes expressionless. He descends the steps and bows—pleasantly.

MELODY

Good morning, Sara.

SARA

Curtly.

Good morning.

Then, ignoring him.

I'm going up and change my dress, Mother.

She goes out right.

Cornelius Melody is forty-five, tall, broad-shouldered, deep-chested, and powerful, with long mus-

cular arms, big feet, and large hairy hands. His heavy-boned body is still firm, erect, and soldierly. Beyond shaky nerves, it shows no effects of hard drinking. It has a bull-like, impervious strength, a tough peasant vitality. It is his face that reveals the ravages of dissipation—a ruined face, which was once extraordinarily handsome in a reckless, arrogant fashion. It is still handsome—the face of an embittered Byronic hero, with a finely chiseled nose over a domineering, sensual mouth set in disdain, pale, hollow-cheeked, framed by thick, curly iron-gray hair. There is a look of wrecked distinction about it, of brooding, humiliated pride. His bloodshot gray eyes have an insulting cold stare which anticipates insult. His manner is that of a polished gentleman. Too much so. He overdoes it and one soon feels that he is overplaying a role which has become more real than his real self to him. But in spite of this, there is something formidable and impressive about him. He is dressed with foppish elegance in old, expensive, finely tailored clothes of the style worn by English aristocracy in Peninsular War days.

MELODY

Advancing into the room—bows formally to his wife.

Good morning, Nora.

His tone condescends. It addresses a person of inferior station.

NORA

Stumbles to her feet—timidly.

Good mornin', Con. I'll get your breakfast.

MELODY

No. Thank you. I want nothing now.

NORA

Coming toward him.

You look pale. Are you sick, Con, darlin'?

MELODY

No.

NORA

Puts a timid hand on his arm.

Come and sit down.

He moves his arm away with instinctive revulsion and goes to the table at center front, and sits in the chair she had occupied. Nora hovers round him.

I'll wet a cloth in cold water to put round your head.

MELODY

No! I desire nothing—except a little peace in which to read the news.

He picks up the paper and holds it so it hides his face from her.

NORA

Meekly.

I'll lave you in peace.

She starts to go to the door at right but turns to stare at him worriedly again. Keeping the paper before his face with his left hand, he reaches out with his right and pours a glass of water from the carafe on the table. Although he cannot see his wife, he is nervously conscious of her. His hand trembles so violently that when he attempts to raise the glass to his lips the water sloshes over his hand and he sets the glass back on the table with a bang. He lowers the paper and explodes nervously.

MELODY

For God's sake, stop your staring!

NORA

I– I was only thinkin' you'd feel better if you'd a bit av food in you.

MELODY

I told you once– !

Controlling his temper.

I am not hungry, Nora.

He raises the paper again. She sighs, her hands fiddling with her apron. A pause.

NORA

Dully.

Maybe it's a hair av the dog you're needin'.

MELODY

As if this were something he had been waiting to hear, his expression loses some of its nervous strain. But he replies virtuously.

No, damn the liquor. Upon my conscience, I've about made up my mind I'll have no more of it. Besides, it's a bit early in the day.

NORA

If it'll give you an appetite–

MELODY

To tell the truth, my stomach is out of sorts.

He licks his lips.

Perhaps a drop wouldn't come amiss.

Nora gets the decanter and glass from the cupboard and sets them before him. She stands gazing at him with a resigned sadness. Melody, his eyes on the paper, is again acutely conscious of her. His nerves cannot stand it. He throws his paper down and bursts out in bitter anger.

Well? I know what you're thinking! Why haven't you the courage to say it for once? By God, I'd have more respect for you! I hate the damned meek of this earth! By the rock of Cashel, I sometimes believe you have always deliberately encouraged me to— It's the one point of superiority you can lay claim to, isn't it?

NORA

Bewilderedly—on the verge of tears.

I don't— It's only your comfort— I can't bear to see you—

MELODY

His expression changes and a look of real affection comes into his eyes. He reaches out a shaking hand to pat her shoulder with an odd, guilty tenderness. He says quietly and with genuine contrition.

Forgive me, Nora. That was unpardonable.

Her face lights up. Abruptly he is ashamed of being ashamed. He looks away and grabs the decanter. Despite his trembling hand he manages to pour a drink and get it to his mouth and drain it. Then he sinks back in his chair and stares at the table, waiting for the liquor to take effect. After a pause he sighs with relief.

I confess I needed that as medicine. I begin to feel more myself.

He pours out another big drink and this time his hand is steadier, and he downs it without much difficulty. He smacks his lips.

By the Immortal, I may have sunk to keeping an inn but at least I've a conscience in my trade. I keep liquor a gentleman can drink.

He starts looking over the paper again—scowls at something—disdainfully, emphasizing his misquote of the line from Byron.

"There shall he rot—Ambition's *dis*honored fool!" The paper is full of the latest swindling lies of that idol of the riffraff,

Andrew Jackson. Contemptible, drunken scoundrel! But he will be the next President, I predict, for all we others can do to prevent. There is a cursed destiny in these decadent times. Everywhere the scum rises to the top.

His eyes fasten on the date and suddenly he strikes the table with his fist.

Today is the 27th! By God, and I would have forgotten!

NORA

Forgot what?

MELODY

The anniversary of Talavera!

NORA

Hastily.

Oh, ain't I stupid not to remember.

MELODY

Bitterly.

I had forgotten myself and no wonder. It's a far cry from this dunghill on which I rot to that glorious day when the Duke of Wellington—Lord Wellesley, then—did me the honor before all the army to commend my bravery.

He glances around the room with loathing.

A far cry, indeed! It would be better to forget!

NORA

Rallying him.

No, no, you mustn't. You've never missed celebratin' it and you won't today. I'll have a special dinner for you like I've always had.

MELODY

With a quick change of manner—eagerly.

Good, Nora. I'll invite Jamie Cregan. It's a stroke of fortune he is here. He served under me at Talavera, as you know. A brave soldier, if he isn't a gentleman. You can place him on my right hand. And we'll have Patch Riley to make music, and O'Dowd and Roche. If they are rabble, they're full of droll humor at times. But put them over there.

He points to the table at left front.

I may tolerate their presence out of charity, but I'll not sink to dining at the same table.

NORA

I'll get your uniform from the trunk, and you'll wear it for dinner like you've done each year.

MELODY

Yes, I must confess I still welcome an excuse to wear it. It makes me feel at least the ghost of the man I was then.

NORA

You're so handsome in it still, no woman could take her eyes off you.

MELODY

With a pleased smile.

I'm afraid you've blarney on your tongue this morning, Nora.

Then boastfully.

But it's true, in those days in Portugal and Spain—

He stops a little shamefacedly, but Nora gives no sign of offense. He takes her hand and pats it gently—avoiding her eyes.

You have the kindest heart in the world, Nora. And I—

His voice breaks.

NORA

Instantly on the verge of grateful tears.

Ah, who wouldn't, Con darlin', when you—

She brushes a hand across her eyes—hastily.

I'll go to the store and get something tasty.

Her face drops as she remembers.

But, God help us, where's the money?

MELODY

Stiffens—haughtily.

Money? Since when has my credit not been good?

NORA

Hurriedly.

Don't fret, now. I'll manage.

He returns to his newspaper, disdaining further interest in money matters.

MELODY

Ha. I see work on the railroad at Baltimore is progressing.

Lowering his paper.

By the Eternal, if I had not been a credulous gull and let the thieving Yankees swindle me of all I had when we came here, that's how I would invest my funds now. And I'd become rich. This country, with its immense territory cannot depend solely on creeping canal boats, as shortsighted fools would have us believe. We must have railroads. Then you will see how quickly America will become rich and great!

His expression changes to one of bitter hatred.

Great enough to crush England in the next war between them, which I know is inevitable! Would I could live to celebrate that victory! If I have one regret for the past—and there are few things in it that do not call for bitter regret—it is that I shed my blood for a country that thanked me with disgrace. But I will be avenged. This country—my country, now—will drive the English from the face of the earth their shameless perfidy has dishonored!

NORA

Glory be to God for that! And we'll free Ireland!

MELODY

Contemptuously.

Ireland? What benefit would freedom be to her unless she could be freed from the Irish?

Then irritably.

But why do I discuss such things with you?

NORA

Humbly.

I know. I'm ignorant.

MELODY

Yet I tried my best to educate you, after we came to America —until I saw it was hopeless.

NORA

You did, surely. And I tried, too, but—

MELODY

You won't even cure yourself of that damned peasant's brogue. And your daughter is becoming as bad.

NORA

She only puts on the brogue to tease you. She can speak as fine as any lady in the land if she wants.

MELODY

Is not listening—sunk in bitter brooding.

But, in God's name, who am I to reproach anyone with anything? Why don't you tell me to examine my own conduct?

NORA

You know I'd never.

MELODY

Stares at her—again he is moved—quietly.

No. I know you would not, Nora.

He looks away—after a pause.

I owe you an apology for what happened last night.

NORA

Don't think of it.

MELODY

With assumed casualness.

Faith, I'd a drink too many, talking over old times with Jamie Cregan.

NORA

I know.

MELODY

I am afraid I may have– The thought of old times– I become bitter. But you understand, it was the liquor talking, if I said anything to wound you.

NORA

I know it.

MELODY

Deeply moved, puts his arm around her.

You're a sweet, kind woman, Nora–too kind.

He kisses her.

NORA

With blissful happiness.

Ah, Con darlin', what do I care what you say when the black thoughts are on you? Sure, don't you know I love you?

MELODY

A sudden revulsion of feeling convulses his face. He bursts out with disgust, pushing her away from him.

For God's sake, why don't you wash your hair? It turns my stomach with its stink of onions and stew!

He reaches for the decanter and shakingly pours a drink. Nora looks as if he had struck her.

NORA

Dully.

I do be washin' it often to plaze you. But when you're standin' over the stove all day, you can't help–

MELODY

Forgive me, Nora. Forget I said that. My nerves are on edge. You'd better leave me alone.

NORA

Her face brightening a little.

Will you ate your breakfast now? I've fine fresh eggs—

MELODY

Grasping at this chance to get rid of her—impatiently.

Yes! In a while. Fifteen minutes, say. But leave me alone now.

She goes out right. Melody drains his drink. Then he gets up and paces back and forth, his hands clasped behind him. The third drink begins to work and his face becomes arrogantly self-assured. He catches his reflection in the mirror on the wall at left and stops before it. He brushes a sleeve fastidiously, adjusts the set of his coat, and surveys himself.

Thank God, I still bear the unmistakable stamp of an officer and a gentleman. And so I will remain to the end, in spite of all fate can do to crush my spirit!

He squares his shoulders defiantly. He stares into his eyes in the glass and recites from Byron's "Childe Harold," as if it were an incantation by which he summons pride to justify his life to himself.

"I have not loved the World, nor the World me;
I have not flattered its rank breath, nor bowed
To its idolatries a patient knee,
Nor coined my cheek to smiles,—nor cried aloud
In worship of an echo: in the crowd
They could not deem me one of such—I stood
Among them, but not of them . . ."

He pauses, then repeats:

"Among them, but not of them." By the Eternal, that

expresses it! Thank God for you, Lord Byron—poet and nobleman who made of his disdain immortal music!

Sara appears in the doorway at right. She has changed to her Sunday dress, a becoming blue that brings out the color of her eyes. She draws back for a moment—then stands watching him contemptuously. Melody senses her presence. He starts and turns quickly away from the mirror. For a second his expression is guilty and confused, but he immediately assumes an air of gentlemanly urbanity and bows to her.

Ah, it's you, my dear. Are you going for a morning stroll? You've a beautiful day for it. It will bring fresh roses to your cheeks.

SARA

I don't know about roses, but it will bring a blush of shame to my cheeks. I have to beg Neilan to give us another month's credit, because you made Mother pay the feed bill for your fine thoroughbred mare!

He gives no sign he hears this. She adds scathingly.

I hope you saw something in the mirror you could admire!

MELODY

In a light tone.

Faith, I suppose I must have looked a vain peacock, preening himself, but you can blame the bad light in my room. One cannot make a decent toilet in that dingy hole in the wall.

SARA

You have the best room in the house, that we ought to rent to guests.

MELODY

Oh, I've no complaints. I was merely explaining my seeming vanity.

SARA

Seeming!

MELODY

Keeping his tone light.

Faith, Sara, you must have risen the wrong side of the bed this morning, but it takes two to make a quarrel and I don't feel quarrelsome. Quite the contrary. I was about to tell you how exceedingly charming and pretty you look, my dear.

SARA

With a mocking, awkward, servant's curtsy—in broad brogue.

Oh, thank ye, yer Honor.

MELODY

Every day you resemble your mother more, as she looked when I first knew her.

SARA

Musha, but it's you have the blarneyin' tongue, God forgive you!

MELODY

In spite of himself, this gets under his skin—angrily.

Be quiet! How dare you talk to me like a common, ignorant—You're my daughter, damn you.

He controls himself and forces a laugh.

A fair hit! You're a great tease, Sara. I shouldn't let you score so easily. Your mother warned me you only did it to provoke me.

Unconsciously he reaches out for the decanter on the table—then pulls his hand back.

SARA

Contemptuously—without brogue now.

Go on and drink. Surely you're not ashamed before me, after all these years.

MELODY

Haughtily.

Ashamed? I don't understand you. A gentleman drinks as he pleases—provided he can hold his liquor as he should.

SARA

A gentleman!

MELODY

Pleasantly again.

I hesitated because I had made a good resolve to be abstemious today. But if you insist—

He pours a drink—a small one—his hand quite steady now.

To your happiness, my dear.

She stares at him scornfully. He goes on graciously.

Will you do me the favor to sit down? I have wanted a quiet chat with you for some time.

He holds out a chair for her at rear of the table at center.

SARA

Eyes him suspiciously—then sits down.

What is it you want?

MELODY

With a playfully paternal manner.

Your happiness, my dear, and what I wish to discuss means happiness to you, unless I have grown blind. How is our patient, young Simon Harford, this morning?

SARA

Curtly.

He's better.

MELODY

I am delighted to hear it.

Gallantly.

How could he help but be with such a charming nurse?

She stares at him coldly. He goes on.

Let us be frank. Young Simon is in love with you. I can see that with half an eye—and, of course, you know it. And you return his love, I surmise.

SARA

Surmise whatever you please.

MELODY

Meaning you do love him? I am glad, Sara.

He becomes sentimentally romantic.

Requited love is the greatest blessing life can bestow on us poor mortals; and first love is the most blessed of all. As Lord Byron has it:

He recites.

"But sweeter still than this, than these, than all,
Is first and passionate Love—it stands alone,
Like Adam's recollection of his fall . . ."

SARA

Interrupts him rudely.

Was it to listen to you recite Byron— ?

MELODY

Concealing discomfiture and resentment—pleasantly.

No. What I was leading up to is that you have my blessing, if that means anything to you. Young Harford is, I am convinced, an estimable youth. I have enjoyed my talks with him. It has been a privilege to be able to converse with a cultured gentleman again. True, he is a bit on the sober side for one so young, but by way of compensation, there is a romantic touch of the poet behind his Yankee phlegm.

SARA

It's fine you approve of him!

MELODY

In your interest I have had some enquiries made about his family.

SARA

Angered—with taunting brogue.

Have you, indade? Musha, that's cute av you! Was it auld Patch Riley, the Piper, made them? Or was it Dan Roche or Paddy O'Dowd, or some other drunken sponge—

MELODY

As if he hadn't heard—condescendingly.

I find his people will pass muster.

SARA

Oh, do you? That's nice!

MELODY

Apparently, his father is a gentleman—that is, by Yankee standards, insofar as one in trade can lay claim to the title. But as I've become an American citizen myself, I suppose it would be downright snobbery to hold to old-world standards.

SARA

Yes, wouldn't it be!

MELODY

Though it is difficult at times for my pride to remember I am no longer the master of Melody Castle and an estate of three thousand acres of as fine pasture and woodlands as you'd find in the whole United Kingdom, with my stable of hunters, and—

SARA

Bitterly.

Well, you've a beautiful thoroughbred mare now, at least—to prove you're still a gentleman!

MELODY

Stung into defiant anger.

Yes, I've the mare! And by God, I'll keep her if I have to starve myself so she may eat.

SARA

You mean, make Mother slave to keep her for you, even if she has to starve!

MELODY

Controls his anger—and ignores this.

But what was I saying? Oh, yes, young Simon's family. His father will pass muster, but it's through his mother, I believe, he comes by his really good blood. My information is, she springs from generations of well-bred gentlefolk.

SARA

It would be a great pride to her, I'm sure, to know you found her suitable!

MELODY

I suppose I may expect the young man to request an interview with me as soon as he is up and about again?

SARA

To declare his honorable intentions and ask you for my hand, is that what you mean?

MELODY

Naturally. He is a man of honor. And there are certain financial arrangements Simon's father or his legal representative will wish to discuss with me. The amount of your settlement has to be agreed upon.

SARA

Stares at him as if she could not believe her ears.

My settlement! Simon's father! God pity you— !

MELODY

Firmly.

Your settlement, certainly. You did not think, I hope, that I would give you away without a penny to your name as if you were some poverty-stricken peasant's daughter? Please remember I have my own position to maintain. Of course, it is a bit difficult at present. I am temporarily hard pressed. But perhaps a mortgage on the inn—

SARA

It's mortgaged to the hilt already, as you very well know.

MELODY

If nothing else, I can always give my note at hand for whatever amount—

SARA

You can give it, sure enough! But who'll take it?

MELODY

Between gentlemen, these matters can always be arranged.

SARA

God help you, it must be a wonderful thing to live in a fairy tale where only dreams are real to you.

Then sharply.

But you needn't waste your dreams worrying about my affairs. I'll thank you not to interfere. Attend to your drinking and leave me alone.

He gives no indication that he has heard a word she has said. She stares at him and a look almost of fear comes into her eyes. She bursts out with a bitter exasperation in which there is a strong undercurrent of entreaty.

Father! Will you never let yourself wake up—not even now when you're sober, or nearly? Is it stark mad you've gone, so you can't tell any more what's dead and a lie, and what's the living truth?

MELODY

His face is convulsed by a spasm of pain as if something vital had been stabbed in him—with a cry of tortured appeal.

Sara!

But instantly his pain is transformed into rage. He half rises from his chair threateningly.

Be quiet, damn you! How dare you— !

She shrinks away and rises to her feet. He forces control on himself and sinks back in his chair, his hands gripping the arms.

The street door at rear is flung open and Dan Roche, Paddy O'Dowd, and Patch Riley attempt to pile in together and get jammed for a moment in the doorway. They all have hangovers, and Roche is talking boisterously. Dan Roche is middle-aged, squat, bowlegged, with a potbelly and short arms lumpy with muscle. His face is flat with a big mouth, protruding ears, and red-rimmed little pig's eyes. He is dressed in dirty, patched clothes. Paddy O'Dowd is thin, round-shouldered, flat-chested, with a pimply complexion, bulgy eyes, and a droopy mouth. His manner is oily and fawning, that of a born sponger and parasite. His clothes are those of a cheap sport. Patch Riley is an old man with a thatch of dirty white hair. His washed-out blue eyes have a wandering, half-witted expression. His skinny body is clothed in rags and there is nothing under his tattered coat but his bare skin. His mouth is sunken in, toothless. He carries an Irish bagpipe under his arm.

ROCHE

His back is half turned as he harangues O'Dowd and Riley, and he does not see Melody and Sara.

And I says, it's Andy Jackson will put you in your place, and all the slave-drivin' Yankee skinflints like you! Take your damned job, I says, and—

O'DOWD

Warningly, his eyes on Melody.

Whist! Whist! Hold your prate!

Roche whirls around to face Melody, and his aggressiveness oozes from him, changing to a hangdog apprehension. For Melody has sprung to his feet, his eyes blazing with an anger which is increased by the glance of contempt Sara casts from him to the three men. O'Dowd avoids Melody's eyes, busies himself in closing the door. Patch Riley stands gazing at Sara with a dreamy, admiring look, lost in a world of his own fancy, oblivious to what is going on.

ROCHE

Placatingly.

Good mornin' to ye, Major.

O'DOWD

Fawning.

Good mornin', yer Honor.

MELODY

How dare you come tramping in here in that manner! Have you mistaken this inn for the sort of dirty shebeen you were used to in the old country where the pigs ran in and out the door?

O'DOWD

We ask pardon, yer Honor.

MELODY

To Roche—an impressive menace in his tone.

You, Paddy. Didn't I forbid you ever to mention that scoundrel Jackson's name in my house or I'd horsewhip the hide off your back?

He takes a threatening step toward him.

Perhaps you think I cannot carry out that threat.

ROCHE

Backs away frightenedly.

No, no, Major. I forgot— Good mornin' to ye, Miss.

O'DOWD

Good mornin', Miss Sara.

She ignores them. Patch Riley is still gazing at her with dreamy admiration, having heard nothing, his hat still on his head. O'Dowd officiously snatches it off for him—rebukingly.

Where's your wits, Patch? Didn't ye hear his Honor?

RILEY

Unheeding—addresses Sara.

Sure it's you, God bless you, looks like a fairy princess as beautiful as a rose in the mornin' dew. I'll raise a tune for you.

He starts to arrange his pipes.

SARA

Curtly.

I want none of your tunes.

Then, seeing the look of wondering hurt in the old man's eyes, she adds kindly.

That's sweet of you, Patch. I know you'd raise a beautiful tune, but I have to go out.

Consoled, the old man smiles at her gratefully.

MELODY

Into the bar, all of you, where you belong! I told you not to use this entrance!

With disdainful tolerance.

I suppose it's a free drink you're after. Well, no one can say of me that I turned away anyone I knew thirsty from my door.

O'DOWD

Thank ye, yer Honor. Come along, Dan.

He takes Riley's arm.

Come on, Patch.

The three go into the bar and O'Dowd closes the door behind them.

SARA

In derisive brogue.

Sure, it's well trained you've got the poor retainers on your American estate to respect the master!

Then as he ignores her and casts a furtive glance at the door to the bar, running his tongue over his dry lips, she says acidly, with no trace of brogue.

Don't let me keep you from joining the gentlemen!

She turns her back on him and goes out the street door at rear.

MELODY

His face is again convulsed by a spasm of pain—pleadingly.

Sara!

Nora enters from the hall at right, carrying a tray with toast, eggs, bacon, and tea. She arranges his breakfast on the table at front center, bustling garrulously.

NORA

Have I kept you waitin'? The divil was in the toast. One lot burned black as a naygur when my back was turned. But the bacon is crisp, and the eggs not too soft, the way you like them. Come and sit down now.

Melody does not seem to hear her. She looks at him worriedly.

What's up with you, Con? Don't you hear me?

O'DOWD

Pokes his head in the door from the bar.

Mickey won't believe you said we could have a drink, yer Honor, unless ye tell him.

MELODY

Licking his lips.

I'm coming.

He goes to the bar door.

NORA

Con! Have this in your stomach first! It'll all get cauld.

MELODY

Without turning to her—in his condescendingly polite tone.

I find I am not the least hungry, Nora. I regret your having gone to so much trouble.

He goes into the bar, closing the door behind him. Nora slumps on a chair at the rear of the table and stares at the breakfast with a pitiful helplessness. She begins to sob quietly.

CURTAIN

Act Two

SCENE *Same as Act One. About half an hour has elapsed. The barroom door opens and Melody comes in. He has had two more drinks and still no breakfast, but this has had no outward effect except that his face is paler and his manner more disdainful. He turns to give orders to the spongers in the bar.*

MELODY

Remember what I say. None of your loud brawling. And you, Riley, keep your bagpipe silent, or out you go. I wish to be alone in quiet for a while with my memories. When Corporal Cregan returns, Mickey, send him in to me. He, at least, knows Talavera is not the name of a new brand of whiskey.

He shuts the door contemptuously on Mickey's "Yes, Major" and the obedient murmur of the others. He sits at rear of the table at left front. At first, he poses to himself, striking an attitude—a Byronic hero, noble, embittered, disdainful, defying his tragic fate, brooding over past glories. But he has no audience and he cannot keep it up. His shoulders sag and he stares at the table top, hopelessness and defeat bringing a trace of real tragedy to his ruined, handsome face.

The street door is opened and Sara enters. He does not hear the click of the latch, or notice her as she comes forward. Fresh from the humiliation of cajoling the storekeeper to extend more credit, her eyes are bitter. At sight of her father they become more so. She moves toward the door at right, determined to ignore him, but something unusual in his attitude strikes her and she stops to regard him searchingly. She starts to say something bitter—stops—finally, in spite of herself, she asks with a trace of genuine pity in her voice.

SARA

What's wrong with you, Father? Are you really sick or is it just—

He starts guiltily, ashamed of being caught in such a weak mood.

MELODY

Gets to his feet politely and bows.

I beg your pardon, my dear. I did not hear you come in.

With a deprecating smile.

Faith, I was far away in spirit, lost in memories of a glorious battle in Spain, nineteen years ago today.

SARA

Her face hardens.

Oh. It's the anniversary of Talavera, is it? Well, I know what that means—a great day for the spongers and a bad day for this inn!

MELODY

Coldly.

I don't understand you. Of course I shall honor the occasion.

SARA

You needn't tell me. I remember the other celebrations—and this year, now Jamie Cregan has appeared, you've an excuse to make it worse.

MELODY

Naturally, an old comrade in arms will be doubly welcome—

SARA

Well, I'll say this much. From the little I've seen of him, I'd rather have free whiskey go down his gullet than the others'. He's a relation, too.

MELODY

Stiffly.

Merely a distant cousin. That has no bearing. It's because Corporal Cregan fought by my side—

SARA

I suppose you've given orders to poor Mother to cook a grand feast for you, as usual, and you'll wear your beautiful uniform, and I'll have the honor of waiting on table. Well, I'll do it just this once more for Mother's sake, or she'd have to, but it'll be the last time.

She turns her back on him and goes to the door at right.

You'll be pleased to learn your daughter had almost to beg on her knees to Neilan before he'd let us have another month's credit. He made it plain it was to Mother he gave it because he pities her for the husband she's got. But what do you care about that, as long as you and your fine thoroughbred mare can live in style!

Melody is shaken for a second. He glances toward the bar as if he longed to return there to escape her. Then he gets hold of himself. His face becomes expressionless. He sits in the same chair and picks up the paper, ignoring her. She starts to go out just as her mother appears in the doorway. Nora is carrying a glass of milk.

NORA

Here's the milk the doctor ordered for the young gentleman. It's time for it, and I knew you'd be going upstairs.

SARA

Takes the milk.

Thank you, Mother.

She nods scornfully toward her father.

I've just been telling him I begged another month's credit from Neilan, so he needn't worry.

NORA

Ah, thank God for that. Neilan's a kind man.

MELODY

Explodes.

Damn his kindness! By the Eternal, if he'd refused, I'd have—!

He controls himself, meeting Sara's contemptuous eyes. He goes on quietly, a bitter, sneering antagonism underneath.

Don't let me detain you, my dear. Take his milk to our Yankee guest, as your mother suggests. Don't miss any chance to play the ministering angel.

Vindictively.

Faith, the poor young devil hasn't a chance to escape with you two scheming peasants laying snares to trap him!

SARA

That's a lie! And leave Mother out of your insults!

MELODY

And if all other tricks fail, there's always one last trick to get him through his honor!

SARA

Tensely.

What trick do you mean?

Nora grabs her arm.

NORA

Hould your prate, now! Why can't you leave him be? It's your fault, for provoking him.

SARA

Quietly.

All right, Mother. I'll leave him to look in the mirror, like he loves to, and remember what he said, and be proud of himself.

Melody winces. Sara goes out right.

MELODY

After a pause—shakenly.

I— She mistook my meaning— It's as you said. She goads me into losing my temper, and I say things—

NORA

Sadly.

I know what made you say it. You think maybe she's like I was, and you can't help remembering my sin with you.

MELODY

Guiltily vehement.

No! No! I tell you she mistook my meaning, and now you—

Then exasperatedly.

Damn your priests' prating about your sin!

With a strange, scornful vanity.

To hear you tell it, you'd think it was you who seduced me! That's likely, isn't it? —remembering the man I was then!

NORA

I remember well. Sure, you was that handsome, no woman could resist you. And you are still.

MELODY

Pleased.

None of your blarney, Nora.

With Byronic gloom.

I am but a ghost haunting a ruin.

Then gallantly but without looking at her.

And how about you in those days? Weren't you the prettiest girl in all Ireland?

Scornfully.

And be damned to your lying, pious shame! You had no shame then, I remember. It was love and joy and glory in you and you were proud!

NORA

Her eyes shining.

I'm still proud and will be to the day I die!

MELODY

Gives her an approving look which turns to distaste at her appearance—looks away irritably.

Why do you bring up the past? I do not wish to discuss it.

NORA

After a pause—timidly.

All the same, you shouldn't talk to Sara as if you thought she'd be up to anything to catch young Harford.

MELODY

I did not think that! She is my daughter—

NORA

She is surely. And he's a dacent lad.

She smiles a bit scornfully.

Sure, from all she's told me, he's that shy he's never dared even to kiss her hand!

MELODY

With more than a little contempt.

I can well believe it. When it comes to making love the Yankees are clumsy, fish-blooded louts. They lack savoir-faire. They have no romantic fire! They know nothing of women.

He snorts disdainfully.

By the Eternal, when I was his age—

Then quickly.

Not that I don't approve of young Harford, mind you. He is a gentleman. When he asks me for Sara's hand I will gladly give my consent, provided his father and I can agree on the amount of her settlement.

NORA

Hastily.

Ah, there's no need to think of that yet.

Then lapsing into her own dream.

Yes, she'll be happy because she loves him dearly, a lot more than she admits. And it'll give her a chance to rise in the world. We'll see the day when she'll live in a grand mansion, dressed in silks and satins, and riding in a carriage with coachman and footman.

MELODY

I desire that as much as you do, Nora. I'm done—finished—no future but the past. But my daughter has the looks, the brains—ambition, youth— She can go far.

Then sneeringly.

That is, if she can remember she's a gentlewoman and stop acting like a bogtrotting peasant wench!

He hears Sara returning downstairs.

She's coming back.

He gets up—bitterly.

As the sight of me seems to irritate her, I'll go in the bar a while. I've had my fill of her insults for one morning.

He opens the bar door. There is a chorus of eager, thirsty welcome from inside. He goes in, closing the door. Sara enters from right. Her face is flushed and her eyes full of dreamy happiness.

NORA

Rebukingly.

Himself went in the bar to be out of reach of your tongue. A fine thing! Aren't you ashamed you haven't enough feeling not to torment him, when you know it's the anniversary—

SARA

All right, Mother. Let him take what joy he can out of the day. I'll even help you get his uniform out of the trunk in the attic and brush and clean it for you.

NORA

Ah, God bless you, that's the way—

Then, astonished at this unexpected docility.

Glory be, but you've changed all of a sudden. What's happened to you?

SARA

I'm so happy now—I can't feel bitter against anyone.

She hesitates—then shyly.

Simon kissed me.

Having said this, she goes on triumphantly.

He got his courage up at last, but it was me made him. I was freshening up his pillows and leaning over him, and he couldn't help it, if he was human.

She laughs tenderly.

And then you'd have laughed to see him. He near sank through the bed with shame at his boldness. He began apologizing as if he was afraid I'd be so insulted I'd never speak to him again.

NORA

Teasingly.

And what did you do? I'll wager you wasn't as brazen as you pretend.

SARA

Ruefully.

It's true, Mother. He made me as bashful as he was. I felt a great fool.

NORA

And was that all? Sure, kissing is easy. Didn't he ask you if you'd marry— ?

SARA

No.

Quickly.

But it was my fault he didn't. He was trying to be brave enough. All he needed was a word of encouragement. But I stood there, dumb as a calf, and when I did speak it was to say I had to come and help you, and the end was I ran from the room, blushing as red as a beet—

She comes to her mother. Nora puts her arm around her. Sara hides her face on her shoulder, on the verge of tears.

Oh, Mother, ain't it crazy to be such a fool?

NORA

Well, when you're in love—

SARA

Breaking away from her—angrily.

That's just it! I'm too much in love and I don't want to be! I won't let my heart rule my head and make a slave of me!

Suddenly she smiles confidently.

Ah well, he loves me as much, and more, I know that, and the next time I'll keep my wits.

She laughs happily.

You can consider it as good as done, Mother. I'm Mrs. Simon Harford, at your pleasure.

She makes a sweeping bow.

NORA

Smiling.

Arrah, none of your airs and graces with me! Help me, now, like you promised, and we'll get your father's uniform out of the trunk. It won't break your back in the attic, like it does me.

SARA

Gaily puts her arm around her mother's waist.

Come along then.

NORA

As they go out right.

I disremember which trunk—and you'll have to help me find the key.

There is a pause. Then the bar door is opened and Melody enters again in the same manner as he did at the beginning of the act. There is the same sound of voices from the bar but this time Melody gives no parting orders but simply shuts the door behind him. He scowls with disgust.

MELODY

Cursed ignorant cattle.

Then with a real, lonely yearning.

I wish Jamie Cregan would come.

Bitterly.

Driven from pillar to post in my own home! Everywhere ignorance—or the scorn of my own daughter!

Then defiantly.

But by the Eternal God, no power on earth, nor in hell itself, can break me!

His eyes are drawn irresistibly to the mirror. He moves in front of it, seeking the satisfying reassurance of his reflection there. What follows is an exact repetition of his scene before the mirror in Act One. There is the same squaring of his shoulders, arrogant lifting of his head, and then the favorite quote from Byron, recited aloud to his own image.

"I have not loved the World, nor the World me;
I have not flattered its rank breath, nor bowed
To its idolatries a patient knee,
Nor coined my cheek to smiles,—nor cried aloud
In worship of an echo: in the crowd
They could not deem me one of such—I stood
Among them, but not of them . . ."

He stands staring in the mirror and does not hear the latch of the street door click. The door opens and Deborah (Mrs. Henry Harford), Simon's mother, enters, closing the door quietly behind her.

Melody continues to be too absorbed to notice anything. For a moment, blinded by the sudden change from the bright glare of the street, she does not see him. When she does, she stares incredulously. Then she smiles with an amused and mocking relish.

Deborah is forty-one, but looks to be no more than thirty. She is small, a little over five feet tall, with a fragile, youthful figure. One would never suspect that she is the middle-aged mother of two grown sons. Her face is beautiful—that is, it is beautiful from the standpoint of the artist with an eye for bone structure and unusual character. It is small, with high cheekbones, wedge-shaped, narrowing from a broad forehead to a square chin, framed by thick, wavy, red-brown hair. The nose is delicate and thin, a trifle aquiline. The mouth, with full lips and even, white teeth, is too large for her face. So are the long-lashed, green-flecked brown eyes, under heavy, angular brows. These would appear large in any face, but in hers they seem enormous and are made more startling by the pallor of her complexion. She has tiny, high-arched feet and thin, tapering hands. Her slender, fragile body is dressed in white with calculated simplicity. About her whole personality is a curious atmosphere of deliberate detachment, the studied aloofness of an ironically amused spectator. Something perversely assertive about it too, as if she consciously carried her originality to the point of whimsical eccentricity.

DEBORAH

I beg your pardon.

Melody jumps and whirls around. For a moment his face has an absurdly startled, stupid look. He is shamed and humiliated and furious at being caught for the second time in one morning before the mirror. His revenge is to draw himself up haughtily and survey her insolently from head to toe. But at once, seeing she is attractive and a lady, his manner changes. Opportunity beckons and he is confident of himself, put upon his mettle. He bows, a gracious, gallant gentleman. There is seductive charm in his welcoming smile and in his voice.

MELODY

Good morning, Mademoiselle. It is an honor to welcome you to this unworthy inn.

He draws out a chair at rear of the larger table in the foreground–bowing again.

If I may presume. You will find it comfortable here, away from the glare of the street.

DEBORAH

Regards him for a second puzzledly. She is impressed in spite of herself by his bearing and distinguished, handsome face.

Thank you.

She comes forward. Melody makes a gallant show of holding her chair and helping her be seated. He takes in all her points with sensual appreciation. It is the same sort of pleasure a lover of horseflesh would have in the appearance of a thoroughbred horse. Meanwhile he speaks with caressing courtesy.

MELODY

Mademoiselle–

He sees her wedding ring.

Pray forgive me, I see it is Madame– Permit me to say again, how great an honor I will esteem it to be of any service.

He manages, as he turns away, as if by accident to brush his hand against her shoulder. She is startled and caught off guard. She shrinks and looks up at him. Their eyes meet and at the nakedly physical appraisement she sees in his, a fascinated fear suddenly seizes her. But at once she is reassured as he shifts his gaze, satisfied by her reactions to his first attack, and hastens to apologize.

I beg your pardon, Madame. I am afraid my manners have grown clumsy with disuse. It is not often a lady comes here now. This inn, like myself, has fallen upon unlucky days.

DEBORAH

Curtly ignoring this.

I presume you are the innkeeper, Melody?

MELODY

A flash of anger in his eyes—arrogantly.

I am *Major* Cornelius Melody, one time of His Majesty's Seventh Dragoons, at your service.

He bows with chill formality.

DEBORAH

Is now an amused spectator again—apologetically.

Oh. Then it is I who owe you an apology, Major Melody.

MELODY

Encouraged—gallantly.

No, no, dear lady, the fault is mine. I should not have taken offense.

With the air of one frankly admitting a praiseworthy weakness.

Faith, I may as well confess my besetting weakness is that of all gentlemen who have known better days. I have a pride unduly sensitive to any fancied slight.

DEBORAH

Playing up to him now.

I assure you, sir, there was no intention on my part to slight you.

MELODY

His eyes again catch hers and hold them—his tone insinuatingly caressing.

You are as gracious as you are beautiful, Madame.

Deborah's amusement is gone. She is again confused and, in spite of herself, frightened and fascinated. Melody proceeds with his attack, full of confidence now, the successful seducer of old. His voice takes on a calculated melancholy cadence. He becomes a romantic, tragic figure, appealing for a woman's understanding and loving compassion.

I am a poor fool, Madame. I would be both wiser and happier if I could reconcile myself to being the proprietor of a tawdry tavern, if I could abjure pride and forget the past. Today of all days it is hard to forget, for it is the anniversary of the battle of Talavera. The most memorable day of my life, Madame. It was on that glorious field I had the honor to be commended for my bravery by the great Duke of Wellington, himself—Sir Arthur Wellesley, then. So I am sure you can find it in your heart to forgive—

His tone more caressing.

One so beautiful must understand the hearts of men full well, since so many must have given their hearts to you.

A coarse passion comes into his voice.

Yes, I'll wager my all against a penny that even among the fish-blooded Yankees there's not a man whose heart doesn't catch flame from your beauty!

He puts his hand over one of her hands on the table and stares into her eyes ardently.

As mine does now!

DEBORAH

Feeling herself borne down weakly by the sheer force of his physical strength, struggles to release her hand. She stammers, with an attempt at lightness.

Is this—what the Irish call blarney, sir?

MELODY

With a fierce, lustful sincerity.

No! I take my oath by the living God, I would charge a square of Napoleon's Old Guard singlehanded for one kiss of your lips.

He bends lower, while his eyes hold hers. For a second it seems he will kiss her and she cannot help herself. Then abruptly the smell of whiskey on his breath brings her to herself, shaken with disgust and coldly angry. She snatches her hand from his and speaks with withering contempt.

DEBORAH

Pah! You reek of whiskey! You are drunk, sir! You are insolent and disgusting! I do not wonder your inn enjoys such meager patronage, if you regale all your guests of my sex with this absurd performance!

Melody straightens up with a jerk, taking a step back as though he had been slapped in the face. Deborah rises to her feet, ignoring him disdainfully. At this moment Sara and her mother enter through the doorway at right. They take in the scene at a glance. Melody and Deborah do not notice their entrance.

NORA

Half under her breath.

Oh, God help us!

SARA

Guesses at once this must be the woman Mickey had told her about. She hurries toward them quickly, trying to hide her apprehension and anger and shame at what she knows must have happened.

What is it, Father? What does the lady wish?

Her arrival is a further blow for Melody, seething now in a fury of humiliated pride. Deborah turns to face Sara.

DEBORAH

Coolly self-possessed—pleasantly.

I came here to see you, Miss Melody, hoping you might know the present whereabouts of my son, Simon.

This is a bombshell for Melody.

MELODY

Blurts out with no apology in his tone but angrily, as if she had intentionally made a fool of him.

You're his mother? In God's name, Madame, why didn't you say so!

DEBORAH

Ignoring him—to Sara.

I've been out to his hermit's cabin, only to find the hermit flown.

SARA

Stammers.

He's here, Mrs. Harford—upstairs in bed. He's been sick—

DEBORAH

Sick? You don't mean seriously?

SARA

Recovering a little from her confusion.

Oh, he's over it now, or almost. It was only a spell of chills and fever he caught from the damp of the lake. I found him there shivering and shaking and made him come here where there's a doctor handy and someone to nurse him.

DEBORAH

Pleasantly.

The someone being you, Miss Melody?

SARA

Yes, me and—my mother and I.

DEBORAH

Graciously.

I am deeply grateful to you and your mother for your kindness.

NORA

Who has remained in the background, now comes forward—with her sweet, friendly smile.

Och, don't be thankin' us, ma'am. Sure, your son is a gentle, fine lad, and we all have great fondness for him. He'd be welcome here if he never paid a penny—

She stops embarrassedly, catching a disapproving glance from Sara. Deborah is repelled by Nora's slovenly appearance, but she feels her simple charm and gentleness, and returns her smile.

SARA

With embarrassed stiffness.

This is my mother, Mrs. Harford.

Deborah inclines her head graciously. Nora instinctively bobs in a peasant's curtsy to one of the gentry. Melody, snubbed and seething, glares at her.

NORA

I'm pleased to make your acquaintance, ma'am.

MELODY

Nora! For the love of God, stop—

Suddenly he is able to become the polished gentleman again—considerately and even a trifle condescendingly.

I am sure Mrs. Harford is waiting to be taken to her son. Am I not right, Madame?

Deborah is so taken aback by his effrontery that for a moment she is speechless. She replies coldly, obviously doing so only because she does not wish to create further embarrassment.

DEBORAH

That is true, sir.

She turns her back on him.

If you will be so kind, Miss Melody. I've wasted so much of the morning and I have to return to the city. I have only time for a short visit—

SARA

Just come with me, Mrs. Harford.

She goes to the door at right, and steps aside to let Deborah precede her.

What a pleasant surprise this will be for Simon. He'd have written you he was sick, but he didn't want to worry you.

She follows Deborah into the hall.

MELODY

Damned fool of a woman! If I'd known— No, be damned if I regret! Cursed Yankee upstart!

With a sneer.

But she didn't fool me with her insulted airs! I've known too many women—

In a rage.

"Absurd performance," was it? God damn her!

NORA

Timidly.

Don't be cursing her and tormenting yourself. She seems a kind lady. She won't hold it against you, when she stops to think, knowing you didn't know who she is.

MELODY

Tensely.

Be quiet!

NORA

Forget it now, do, for Sara's sake. Sure, you wouldn't want anything to come between her and the lad.

He is silent. She goes on comfortingly.

Go on up to your room now and you'll find something to take your mind off. Sara and I have your uniform brushed and laid out on the bed.

MELODY

Harshly.

Put it back in the trunk! I don't want it to remind me—

With humiliated rage again.

By the Eternal, I'll wager she believed what I told her of Talavera and the Great Duke honoring me was a drunken liar's boast!

NORA

No, she'd never, Con. She couldn't.

MELODY

Seized by an idea.

Well, seeing would be believing, eh, my fine lady? Yes, by God, that will prove to her—

He turns to Nora, his self-confidence partly restored.

Thank you for reminding me of my duty to Sara. You are right. I do owe it to her interests to forget my anger and make a formal apology to Simon's mother for our little misunderstanding.

He smiles condescendingly.

Faith, as a gentleman, I should grant it is a pretty woman's privilege to be always right even when she is wrong.

He goes to the door at extreme left front and opens it.

If the lady should come back, kindly keep her here on some excuse until I return.

This is a command. He disappears, closing the door behind him.

NORA

Sighs.

Ah well, it's all right. He'll be on his best behavior now, and he'll feel proud again in his uniform.

She sits at the end of center table right and relaxes wearily. A moment later Sara enters quickly from right and comes to her.

SARA

Where's Father?

NORA

I got him to go up and put on his uniform. It'll console him.

SARA

Bitterly.

Console *him?* It's me ought to be consoled for having such a great fool for a father!

NORA

Hush now! How could he know who— ?

SARA

With a sudden reversal of feeling—almost vindictively.

Yes, it serves her right. I suppose she thinks she's such a great lady anyone in America would pay her respect. Well, she knows better now. And she didn't act as insulted as she might. Maybe she liked it, for all her pretenses.

Again with an abrupt reversal of feeling.

Ah, how can I talk such craziness! Him and his drunken love-making! Well, he got put in his place, and aren't I glad! He won't forget in a hurry how she snubbed him, as if he was no better than dirt under her feet!

NORA

She didn't. She had the sense to see he'd been drinking and not to mind him.

SARA

Dully.

Maybe. But isn't that bad enough? What woman would want her son to marry the daughter of a man like—

She breaks down.

Oh, Mother, I was feeling so happy and sure of Simon, and now— Why did she have to come today? If she'd waited till tomorrow, even, I'd have got him to ask me to marry him, and once he'd done that no power on earth could change him.

NORA

If he loves you no power can change him, anyway.

Proudly.

Don't I know!

Reassuringly.

She's his mother, and she loves him and she'll want him to be happy, and she'll see he loves you. What makes you think she'll try to change him?

SARA

Because she hates me, Mother—for one reason.

NORA

She doesn't. She couldn't.

SARA

She does. Oh, she acted as nice as nice, but she didn't fool me. She's the kind would be polite to the hangman, and her on the scaffold.

She lowers her voice.

It isn't just to pay Simon a visit she came. It's because Simon's father got a letter telling him about us, and he showed it to her.

NORA

Who did a dirty trick like that?

SARA

It wasn't signed, she said. I suppose someone around here that hates Father—and who doesn't?

NORA

Bad luck to the blackguard, whoever it was!

SARA

She said she'd come to warn Simon his father is wild with anger and he's gone to see his lawyer— But that doesn't worry me. It's only her influence I'm afraid of.

NORA

How do you know about the letter?

SARA

Avoiding her eyes.

I sneaked back to listen outside the door.

NORA

Shame on you! You should have more pride!

SARA

I was ashamed, Mother, after a moment or two, and I came away.

Then defiantly.

No, I'm not ashamed. I wanted to learn what tricks she might be up to, so I'll be able to fight them. I'm not ashamed at all. I'll do anything to keep him.

Lowering her voice.

She started talking the second she got in the door. She had only a few minutes because she has to be home before dinner so her husband won't suspect she came here. He's forbidden her to see Simon ever since Simon came out here to live.

NORA

Well, doesn't her coming against her husband's orders show she's on Simon's side?

SARA

Yes, but it doesn't show she wants him to marry me.

Impatiently.

Don't be so simple, Mother. Wouldn't she tell Simon that anyway, even if the truth was her husband sent her to do all she could to get him away from me?

NORA

Don't look for trouble before it comes. Wait and see, now. Maybe you'll find—

SARA

I'll find what I said, Mother—that she hates me.

Bitterly.

Even if she came here with good intentions, she wouldn't have them now, after our great gentleman has insulted her. Thank God, if he's putting on his uniform, he'll be hours before the mirror, and she'll be gone before he can make a fool of himself again.

Nora starts to tell her the truth—then thinks better of it. Sara goes on, changing her tone.

But I'd like her to see him in his uniform, at that, if he was sober. She'd find she couldn't look down on him—

Exasperatedly.

Och! I'm as crazy as he is. As if she hadn't the brains to see through him.

NORA

Wearily.

Leave him be, for the love of God.

SARA

After a pause—defiantly.

Let her try whatever game she likes. I have brains too, she'll discover.

Then uneasily.

Only, like Simon's told me, I feel she's strange and queer behind her lady's airs, and it'll be hard to tell what she's really up to.

They both hear a sound from upstairs.

That's her, now. She didn't waste much time. Well, I'm ready for her. Go in the kitchen, will you, Mother? I want to give her the chance to have it out with me alone.

Nora gets up—then, remembering Melody's orders, glances toward the door at left front uneasily and hesitates. Sara says urgently.

Don't you hear me? Hurry, Mother!

Nora sighs and goes out quickly, right. Sara sits at rear of the center table and waits, drawing herself up in an unconscious imitation of her father's grand manner. Deborah appears in the doorway at right. There is nothing in her expression to betray any emotion resulting from her interview with her son. She smiles pleasantly at Sara, who rises graciously from her chair.

DEBORAH

Coming to her.

I am glad to find you here, Miss Melody. It gives me another opportunity to express my gratitude for your kindness to my son during his illness.

SARA

Thank you, Mrs. Harford. My mother and I have been only too happy to do all we could.

She adds defiantly.

We are very fond of Simon.

DEBOROAH

A glint of secret amusement in her eyes.

Yes, I feel you are. And he has told me how fond he is of you.

Her manner becomes reflective. She speaks rapidly in a remote, detached way, lowering her voice unconsciously as if she were thinking aloud to herself.

This is the first time I have seen Simon since he left home to seek self-emancipation at the breast of Nature. I find him not so greatly changed as I had been led to expect from his letters. Of course, it is some time since he has written. I had thought his implacably honest discovery that the poetry he hoped the pure freedom of Nature would inspire him to write is, after all, but a crude imitation of Lord Byron's would have more bitterly depressed his spirit.

She smiles.

But evidently he has found a new romantic dream by way of recompense. As I might have known he would. Simon is an inveterate dreamer—a weakness he inherited from me, I'm afraid, although I must admit the Harfords have been great dreamers, too, in their way. Even my husband has a dream—a conservative, material dream, naturally. I have just been reminding Simon that his father is rigidly unforgiving when his dream is flouted, and very practical in his methods of defending it.

She smiles again.

My warning was the mechanical gesture of a mother's duty, merely. I realized it would have no effect. He did not listen to what I said. For that matter, neither did I.

She laughs a little detached laugh, as if she were secretly amused.

SARA

Stares at her, unable to decide what is behind all this and how she should react—with an undercurrent of resentment.

I don't think Simon imitates Lord Byron. I hate Lord Byron's poetry. And I know there's a true poet in Simon.

DEBORAH

Vaguely surprised—speaks rapidly again.

Oh, in feeling, of course. It is natural you should admire that in him—now. But I warn you it is a quality difficult for a woman to keep on admiring in a Harford, judging from what I know of the family history. Simon's great-grandfather, Jonathan Harford, had it. He was killed at Bunker Hill, but I suspect the War for Independence was merely a symbolic opportunity for him. His was a personal war, I am sure—for pure freedom. Simon's grandfather, Evan Harford, had the quality too. A fanatic in the cause of pure freedom, he became scornful of our Revolution. It made too many compromises with the ideal to free him. He went to France and became a rabid Jacobin, a worshiper of Robespierre. He would have liked to have gone to the guillotine with his incorruptible Redeemer, but he was too unimportant. They simply forgot to kill him. He came home and lived in a little temple of Liberty he had built in a corner of what is now my garden. It is still there. I remember him well. A dry, gentle, cruel, indomitable, futile old idealist who used frequently to wear his old uniform of the French Republican National Guard. He died wearing it. But the point is, you can have no idea what revengeful hate the Harford pursuit of freedom imposed upon the women who shared their lives. The three daughters-in-law of Jonathan, Evan's half-sisters, had to make a large, greedy fortune out of privateering and the Northwest trade, and finally were even driven to embrace the profits of the slave trade—as a triumphant climax, you understand, of their long battle to escape the enslavement of freedom by enslaving it. Evan's wife, of course, was drawn into this conflict, and became their tool and accomplice. They even attempted to own me, but I managed to escape because there was so little of me in the flesh that aged, greedy fingers could clutch. I am sorry they are dead and cannot know you. They would approve of you, I think. They would see that you are strong and ambitious and determined to take what you want. They would have smiled like senile, hungry serpents and welcomed you into their coils.

She laughs.

Evil old witches! Detestable, but I could not help admiring them—pitying them, too—in the end. We had a bond in common. They idolized Napoleon. They used to say he was the only man they would ever have married. And I used to dream I was Josephine—even after my marriage, I'm afraid. The Sisters, as everyone called them, and all of the family accompanied my husband and me on our honeymoon—to Paris to witness the Emperor's coronation.

She pauses, smiling at her memories.

SARA

Against her will, has become a bit hypnotized by Deborah's rapid, low, musical flow of words, as she strains to grasp the implication for her. She speaks in a low, confidential tone herself, smiling naturally.

I've always admired him too. It's one of the things I've held against my father, that he fought against him and not for him.

DEBORAH

Starts, as if awakening—with a pleasant smile.

Well, Miss Melody, this is tiresome of me to stand here giving you a discourse on Harford family history. I don't know what you must think of me—but doubtless Simon has told you I am a bit eccentric at times.

She glances at Sara's face—amusedly.

Ah, I can see he has. Then I am sure you will make allowances. I really do not know what inspired me—except perhaps, that I wish to be fair and warn you, too.

SARA

Stiffens.

Warn me about what, Mrs. Harford?

DEBORAH

Why, that the Harfords never part with their dreams even when they deny them. They cannot. That is the family curse. For example, this book Simon plans to write to denounce the evil of greed and possessive ambition, and uphold the virtue of freeing oneself from the lust for power and saving our souls by being content with little. I cannot imagine you taking that seriously.

She again flashes a glance at Sara.

I see you do not. Neither do I. I do not even believe Simon will ever write this book on paper. But I warn you it is already written on his conscience and–

She stops with a little disdaining laugh.

I begin to resemble Cassandra with all my warnings. And I continue to stand here boring you with words.

She holds out her hand graciously.

Goodbye, Miss Melody.

SARA

Takes her hand mechanically.

Goodbye, Mrs. Harford.

Deborah starts for the door at rear. Sara follows her, her expression confused, suspicious, and at the same time hopeful. Suddenly she blurts out impulsively.

Mrs. Harford, I–

DEBORAH

Turns on her, pleasantly.

Yes, Miss Melody?

But her eyes have become blank and expressionless and discourage any attempt at further contact.

SARA

Silenced—with stiff politeness.

Isn't there some sort of cooling drink I could get you before you go? You must be parched after walking from the road to Simon's cabin and back on this hot day.

DEBORAH

Nothing, thank you.

Then talking rapidly again in her strange detached way.

Yes, I did find my walk alone in the woods a strangely overpowering experience. Frightening—but intoxicating, too. Such a wild feeling of release and fresh enslavement. I have not ventured from my garden in many years. There, nature is tamed, constrained to obey and adorn. I had forgotten how compelling the brutal power of primitive, possessive nature can be—when suddenly one is attacked by it.

She smiles.

It has been a most confusing morning for a tired, middle-aged matron, but I flatter myself I have preserved a philosophic poise, or should I say, pose, as well as may be. Nevertheless, it will be a relief to return to my garden and book and meditations and listen indifferently again while the footsteps of life pass and recede along the street beyond the high wall. I shall never venture forth again to do my duty. It is a noble occupation, no doubt, for those who can presume they know what their duty to others is; but I—

She laughs.

Mercy, here I am chattering on again.

She turns to the door.

Cato will be provoked at me for keeping him waiting. I've already caused his beloved horses to be half-devoured by flies. Cato is our black coachman. He also is fond of Simon, although since Simon became emancipated he has embarrassed Cato acutely by shaking his hand whenever they meet. Cato was always a self-possessed free man even when he was a slave. It astonishes him that Simon has to prove that he—I mean Simon—is free.

She smiles.

Goodbye again, Miss Melody. This time I really am going.

Sara opens the door for her. She walks past Sara into the street, turns left, and, passing before the two windows, disappears. Sara closes the door and comes back slowly to the head of the table. She stands thinking, her expression puzzled, apprehensive, and resentful. Nora appears in the doorway at right.

NORA

God forgive you, Sara, why did you let her go? Your father told me—

SARA

I can't make her out, Mother. You'd think she didn't care, but she does care. And she hates me. I could feel it. But you can't tell— She's crazy, I think. She talked on and on as if she couldn't stop—queer blather about Simon's ancestors, and herself, and Napoleon, and Nature, and her garden and freedom, and God knows what—but letting me know all the time she had a meaning behind it, and was warning and threatening me. Oh, she may be daft in some ways, but she's no fool. I know she didn't let Simon guess she'd rather have him dead than married to me. Oh, no, I'm sure she told him if he was sure he loved me and I meant his happiness— But then she'd say he ought to wait and prove he's sure—anything to give her time. She'd make him promise to wait. Yes, I'll wager that's what she's done!

NORA

Who has been watching the door at left front, preoccupied by her own worry—frightenedly.

Your father'll be down any second. I'm going out in the garden.

She grabs Sara's arm.

Come along with me, and give him time to get over his rage.

SARA

Shakes off her hand—exasperatedly.

Leave me be, Mother. I've enough to worry me without bothering about him. I've got to plan the best way to act when I see Simon. I've got to be as big a liar as she was. I'll have to pretend I liked her and I'd respect whatever advice she gave him. I mustn't let him see— But I won't go to him again today, Mother. You can take up his meals and his milk, if you will. Tell him I'm too busy. I want to get him anxious and afraid maybe I'm mad at him for something, that maybe his mother said something. If he once has the idea maybe he's lost me—that ought to help, don't you think, Mother?

NORA

Sees the door at left front begin to open—in a whisper.

Oh, God help me!

She turns in panicky flight and disappears through the doorway, right.

The door at left front slowly opens—slowly because Melody, hearing voices in the room and hoping Deborah is there, is deliberately making a dramatic entrance. And in spite of its obviousness, it is effective. Wearing the brilliant scarlet full-dress uniform of a major in one of Wellington's dragoon regiments, he looks extraordinarily handsome and distinguished—a startling, colorful, romantic figure, possessing now a genuine quality he has not had before, the quality of the formidably strong, disdainfully fearless cavalry officer he really had been. The uniform has been preserved with the greatest care. Each button is shining and the cloth is spotless.

Being in it has notably restored his self-confident arrogance. Also, he has done everything he can to freshen up his face and hide any effect of his morning's drinks. When he discovers Deborah is not in the room, he is mildly disappointed and, as always when he first confronts Sara alone, he seems to shrink back guiltily within himself. Sara's face hardens and she gives no sign of knowing he is there. He comes slowly around the table at left front, until he stands at the end of the center table facing her. She still refuses to notice him and he is forced to speak. He does so with the air of one who condescends to be amused by his own foibles.

MELODY

I happened to go to my room and found you and your mother had laid out my uniform so invitingly that I could not resist the temptation to put it on at once instead of waiting until evening.

SARA

Turns on him. In spite of herself she is so struck by his appearance that the contempt is forced back and she can only stammer a bit foolishly.

Yes, I—I see you did.

There is a moment's pause. She stares at him fascinatedly—then blurts out with impulsive admiration.

You look grand and handsome, Father.

MELODY

As pleased as a child.

Why, it is most kind of you to say that, my dear Sara.

Preening himself.

I flatter myself I do not look too unworthy of the man I was when I wore this uniform with honor.

SARA

An appeal forced out of her that is both pleading and a bitter reproach.

Oh, Father, why can't you ever be the thing you can seem to be?

A sad scorn comes into her voice.

The man you were. I'm sorry I never knew that soldier. I think he was the only man who wasn't just a dream.

MELODY

His face becomes a blank disguise—coldly.

I don't understand you.

A pause. He begins to talk in an arrogantly amused tone.

I suspect you are still holding against me my unfortunate blunder with your future mother-in-law. I would not blame you if you did.

He smiles.

Faith, I did put my foot in it.

He chuckles.

The devil of it is, I can never get used to these Yankee ladies. I do them the honor of complimenting them with a bit of harmless flattery and, lo and behold, suddenly my lady acts as if I had insulted her. It must be their damned narrow Puritan background. They can't help seeing sin hiding under every bush, but this one need not have been alarmed. I never had an eye for skinny, pale snips of women—

Hastily.

But what I want to tell you is I am sorry it happened, Sara, and I will do my best, for the sake of your interests, to make honorable amends. I shall do the lady the honor of tendering her my humble apologies when she comes downstairs.

With arrogant vanity.

I flatter myself she will be graciously pleased to make peace.

She was not as outraged by half as her conscience made her pretend, if I am any judge of feminine frailty.

SARA

Who has been staring at him with scorn until he says this last—impulsively, with a sneer of agreement.

I'll wager she wasn't for all her airs.

Then furious at herself and him.

Ah, will you stop telling me your mad dreams!

Controlling herself—coldly.

You'll have no chance to make bad worse by trying to fascinate her with your beautiful uniform. She's gone.

MELODY

Stunned.

Gone?

Furiously.

You're lying, damn you!

SARA

I'm not. She left ten minutes ago, or more.

MELODY

Before he thinks.

But I told your mother to keep her here until—

He stops abruptly.

SARA

So that's why Mother is so frightened. Well, it was me let her go, so don't take out your rage on poor Mother.

MELODY

Rage? My dear Sara, all I feel is relief. Surely you can't believe I could have looked forward to humbling my pride, even though it would have furthered your interests.

SARA

Furthered my interests by giving her another reason to laugh up her sleeve at your pretenses?

With angry scorn, lapsing into broad brogue.

Arrah, God pity you!

She turns her back on him and goes off, right. Melody stands gripping the back of the chair at the foot of the table in his big, powerful hands in an effort to control himself. There is a crack as the chair back snaps in half. He stares at the fragment in his hand with stupid surprise. The door to the bar is shoved open and Mickey calls in.

MALOY

Here's Cregan back to see you, Major.

MELODY

Startled, repeats stupidly.

Cregan?

Then his face suddenly lights up with pathetic eagerness and his voice is full of welcoming warmth as he calls.

Jamie! My old comrade in arms!

As Cregan enters, he grips his hand.

By the Powers, I'm glad you're here, Jamie.

Cregan is surprised and pleased by the warmth of his welcome. Melody draws him into the room.

Come. Sit down. You'll join me in a drink, I know.

He gets Cregan a glass from the cupboard. The decanter and Melody's glass are already on the table.

CREGAN

Admiringly.

Be God, it's the old uniform, no less, and you look as fine a figure in it as ever you did in Spain.

He sits at right of table at left front as Melody sits at rear.

MELODY

Immensely pleased—deprecatingly.

Hardly, Jamie—but not a total ruin yet, I hope. I put it on in honor of the day. I see you've forgotten. For shame, you dog, not to remember Talavera.

CREGAN

Excitedly.

Talavera, is it? Where I got my saber cut. Be the mortal, I remember it, and you've a right to celebrate. You was worth any ten men in the army that day!

Melody has shoved the decanter toward him. He pours a drink.

MELODY

This compliment completely restores him to his arrogant self.

Yes, I think I may say I did acquit myself with honor.

Patronizingly.

So, for that matter, did you.

He pours a drink and raises his glass.

To the day and your good health, Corporal Cregan.

CREGAN

Enthusiastically.

To the day and yourself, God bless you, Con!

He tries to touch brims with Melody's glass, but Melody holds his glass away and draws himself up haughtily.

MELODY

With cold rebuke.

I said, to the day and your good health, *Corporal Cregan.*

CREGAN

For a second is angry—then he grins and mutters admiringly.

Be God, it's you can bate the world and never let it change you!

Correcting his toast with emphasis.

To the day and yourself, *Major Melody.*

MELODY

Touches his glass to Cregan's*—graciously condecending.*

Drink hearty, Corporal.

They drink.

CURTAIN

Act Three

SCENE *The same. The door to the bar is closed. It is around eight that evening and there are candles on the center table. Melody sits at the head of this table. In his brilliant uniform he presents more than ever an impressively colorful figure in the room, which appears smaller and dingier in the candlelight. Cregan is in the chair on his right. The other chairs at this table are unoccupied. Riley, O'Dowd, and Roche sit at the table at left front. Riley is at front, but his chair is turned sideways so he faces right. O'Dowd has the chair against the wall, facing right, with Roche across the table from him, his back to Melody. All five are drunk, Melody more so than any of them, but except for the glazed glitter in his eyes and his deathly pallor, his appearance does not betray him. He is holding his liquor like a gentleman.*

Cregan is the least drunk. O'Dowd and Roche are boisterous. The effect of the drink on Riley is merely to sink him deeper in dreams. He seems oblivious to his surroundings.

An empty and a half-empty bottle of port are on the table before Melody and Cregan, and their glasses are full. The three at the table have a decanter of whiskey.

Sara, wearing her working dress and an apron, is removing dishes and the remains of the dinner. Her face is set. She is determined to ignore them, but there is angry disgust in her eyes. Melody is arranging forks, knives, spoons, saltcellar, etc., in a plan of battle on the table before him. Cregan watches him. Patch Riley gives a few tuning-up quavers on his pipes.

MELODY

Here's the river Tagus. And here, Talavera. This would be the French position on a rise of ground with the plain between our lines and theirs. Here is our redoubt with the Fourth Division and the Guards. And here's our cavalry brigade in a valley toward our left, if you'll remember, Corporal Cregan.

CREGAN

Excitedly.

Remember? Sure I see it as clear as yesterday!

RILEY

Bursts into a rollicking song, accompanying himself on the pipes, his voice the quavering ghost of tenor but still true—to the tune of "Baltiorum."

"She'd a pig and boneens,
She'd a bed and a dresser,
And a nate little room
For the father confessor;
With a cupboard and curtains, and something, I'm towld,
That his riv'rance liked when the weather was cowld.
And it's hurroo, hurroo! Biddy O'Rafferty!"

Roche and O'Dowd roar after him, beating time on the table with their glasses—"Hurroo, hurroo! Biddy O'Rafferty!"*—and laugh drunkenly. Cregan, too, joins in this chorus. Melody frowns angrily at the interruption, but at the end he smiles with lordly condescension, pleased by the irreverence of the song.*

O'DOWD

After a cunning glance at Melody's face to see what his reaction is—derisively.

Och, lave it to the priests, divil mend thim! Ain't it so, Major?

MELODY

Ay, damn them all! A song in the right spirit, Piper. Faith, I'll have you repeat it for my wife's benefit when she joins us. She still has a secret fondness for priests. And now, less noise, you blackguards. Corporal Cregan and I cannot hear each other with your brawling.

O'DOWD

Smirkingly obedient.

Quiet it is, yer Honor. Be quiet, Patch.

He gives the old man, who is lost in dreams, a shove that almost knocks him off his chair. Riley stares at him bewilderedly. O'Dowd and Roche guffaw.

MELODY

Scowls at them, then turns to Cregan.

Where was I, Corporal? Oh, yes, we were waiting in the valley. We heard a trumpet from the French lines and saw them forming for the attack. An aide-de-camp galloped down the hill to us—

SARA

Who has been watching him disdainfully, reaches out to take his plate—rudely in mocking brogue.

I'll have your plate, av ye plaze, Major, before your gallant dragoons charge over it and break it.

MELODY

Holds his plate on the table with one hand so she cannot take it, and raises his glass of wine with the other—ignoring her.

Wet your lips, Corporal. Talavera was a devilish thirsty day, if you'll remember.

He drinks.

CREGAN

Glances uneasily at Sara.

It was that.

He drinks.

MELODY

Smacking his lips.

Good wine, Corporal. Thank God, I still have wine in my cellar fit for a gentleman.

SARA

Angrily.

Are you going to let me take your plate?

MELODY

Ignoring her.

No, I have no need to apologize for the wine. Nor for the dinner, for that matter. Nora is a good cook when she forgets her infernal parsimony and buys food that one can eat without disgust. But I do owe you an apology for the quality of the service. I have tried to teach the waitress not to snatch plates from the table as if she were feeding dogs in a kennel but she cannot learn.

He takes his hand from the plate—to Sara.

There. Now let me see you take it properly.

She stares at him for a moment, speechless with anger—then snatches the plate from in front of him.

CREGAN

Hastily recalls Melody to the battlefield.

You were where the aide-de-camp galloped up to us, Major. It was then the French artillery opened on us.

Sara goes out right, carrying a tray laden with plates.

MELODY

We charged the columns on our left—here—

He marks the tablecloth.

that were pushing back the Guards. I'll never forget the blast of death from the French squares. And then their chasseurs and lancers were on us! By God, it's a miracle any of us came through!

CREGAN

You wasn't touched except you'd a bullet through your coat, but I had this token on my cheek to remember a French saber by.

MELODY

Brave days, those! By the Eternal, then one lived! Then one forgot!

He stops—when he speaks again it is bitterly.

Little did I dream then the disgrace that was to be my reward later on.

CREGAN

Consolingly.

Ah well, that's the bad luck of things. You'd have been made a colonel soon, if you'd left the Spanish woman alone and not fought that duel.

MELODY

Arrogantly threatening.

Are you presuming to question my conduct in that affair, Corporal Cregan?

CREGAN

Hastily.

Sorra a bit! Don't mind me, now.

MELODY

Stiffly.

I accept your apology.

He drinks the rest of his wine, pours another glass, then stares moodily before him. Cregan drains his glass and refills.

O'DOWD

Peering past Roche to watch Melody, leans across to Roche—in a sneering whisper.

Ain't he the lunatic, sittin' like a play-actor in his red coat, lyin' about his battles with the French!

ROCHE

Sullenly—but careful to keep his voice low.

He'd ought to be shamed he ivir wore the bloody red av England, God's curse on him!

O'DOWD

Don't be wishin' him harm, for it's thirsty we'd be without him. Drink long life to him, and may he always be as big a fool as he is this night!

He sloshes whiskey from the decanter into both their glasses.

ROCHE

With a drunken leer.

Thrue for you! I'll toast him on that.

He twists round to face Melody, holds up his glass and bawls.

To the grandest gintleman ivir come from the shores av Ireland! Long life to you, Major!

O'DOWD

Hurroo! Long life, yer Honor!

RILEY

Awakened from his dream, mechanically raises his glass.

And to all that belong to ye.

MELODY

Startled from his thoughts, becomes at once the condescending squire—smiling tolerantly.

I said, less noise, you dogs. All the same, I thank you for your toast.

They drink. A pause. Abruptly Melody begins to recite from Byron. He reads the verse well, quietly, with a bitter eloquence.

"But midst the crowd, the hum, the shock of men,
To hear, to see, to feel, and to possess,
And roam along, the World's tired denizen,
With none who bless us, none whom we can bless;
Minions of Splendour shrinking from distress!
None that, with kindred consciousness endued,
If we were not, would seem to smile the less,
Of all that flattered—followed—sought, and sued;
This is to be alone— This, this is Solitude!"

He stops and glances from one face to another. Their expressions are all blank. He remarks with insulting derisiveness.

What? You do not understand, my lads? Well, all the better for you. So may you go on fooling yourselves that I am fooled in you.

Then with a quick change of mood, heartily.

Give us a hunting song, Patch. You've not forgotten "Modideroo," I'll be bound.

RILEY

Roused to interest immediately.

Does a duck forget wather? I'll show ye!

He begins the preliminary quavers on his pipes.

O'DOWD

Modideroo!

ROCHE

Hurroo!

RILEY.

Accompanying himself, sings with wailing melancholy the first verse that comes to his mind of an old hunting song.

"And the fox set him down and looked about,
And many were feared to follow;
'Maybe I'm wrong,' says he, 'but I doubt
That you'll be as gay tomorrow.
For loud as you cry, and high as you ride,
And little you feel my sorrow,
I'll be free on the mountainside
While you'll lie low tomorrow.'
Oh, Modideroo, aroo, aroo!"

Melody, excited now, beats time on the table with his glass along with Cregan, Roche, and O'Dowd, and all bellow the refrain, "Oh, Modideroo, aroo, aroo!"

MELODY

His eyes alight, forgetting himself, a strong lilt of brogue coming into his voice.

Ah, that brings it back clear as life! Melody Castle in the days that's gone! A wind from the south, and a sky gray with clouds—good weather for the hounds. A true Irish hunter under me that knows and loves me and would raise to a jump over hell if I gave the word! To hell with men, I say!—and women, too!—with their cowardly hearts rotten and stinking with lies and greed and treachery! Give me a horse to love and I'll cry quits to men! And then away, with the hounds in full cry, and after them! Off with divil a care for your neck, over ditches and streams and stone walls and fences, the fox doubling up the mountainside through the furze and the heather— !

Sara has entered from right as he begins this longing invocation of old hunting days. She stands behind his chair, listening contemptuously. He suddenly feels her presence and turns his head. When he catches the sneer in her eyes, it is as if cold water were dashed in his face. He addresses her as if she were a servant.

Well? What is it? What are you waiting for now?

SARA

Roughly, with coarse brogue.

What would I be waitin' for but for you to get through with your blather about lovin' horses, and give me a chance to finish my work? Can't you—and the other gintlemen—finish gettin' drunk in the bar and lave me clear the tables?

O'Dowd conceals a grin behind his hand; Roche stifles a malicious guffaw.

CREGAN

With an apprehensive glance at Melody, shakes his head at her admonishingly.

Now, Sara, be aisy.

But Melody suppresses any angry reaction. He rises to his feet, a bit stiffly and carefully, and bows.

MELODY

Coldly.

I beg your pardon if we have interfered with your duties.

To O'Dowd and his companions.

Into the bar, you louts!

O'DOWD

The bar it is, sorr. Come, Dan. Wake up, Patch.

He pokes the piper. He and Roche go into the bar, and Riley stumbles vaguely after them. Cregan waits for Melody.

MELODY

Go along, Corporal. I'll join you presently. I wish to speak to my daughter.

CREGAN

All right, Major.

He shakes his head at Sara, as if to say, don't provoke him. She ignores him. He goes into the bar, closing the door behind him. She stares at her father with angry disgust.

SARA

You're drunk. If you think I'm going to stay here and listen to—

MELODY

His face expressionless, draws out his chair at the head of the center table for her—politely.

Sit down, my dear.

SARA

I won't. I have no time. Poor Mother is half dead on her feet. I have to help her. There's a pile of dishes to wash after your grand anniversary feast!

With bitter anger.

Thank God it's over, and it's the last time you'll ever take satisfaction in having me wait on table for drunken scum like O'Dowd and—

MELODY

Quietly.

A daughter who takes satisfaction in letting even the scum see that she hates and despises her father!

He shrugs his shoulders.

But no matter.

Indicating the chair again.

Won't you sit down, my dear?

SARA

If you ever dared face the truth, you'd hate and despise yourself!

Passionately.

All I pray to God is that someday when you're admiring yourself in the mirror something will make you see at last what you really are! That will be revenge in full for all you've done to Mother and me!

She waits defiantly, as if expecting him to lose his temper and curse her. But Melody acts as if he had not heard her.

MELODY

His face expressionless, his manner insistently bland and polite.

Sit down, my dear. I will not detain you long, and I think you will find what I have to tell you of great interest.

She searches his face, uneasy now, feeling a threat hidden behind his cold, quiet, gentlemanly tone. She sits down and he sits at rear of table, with an empty chair separating them.

SARA

You'd better think well before you speak, Father. I know the devil that's in you when you're quiet like this with your brain mad with drink.

MELODY

I don't understand you. All I wish is to relate something which happened this afternoon.

SARA

Giving way to bitterness at her humiliation again—sneeringly.

When you went riding on your beautiful thoroughbred mare while Mother and I were sweating and suffocating in the heat of the kitchen to prepare your Lordship's banquet? Sure, I hope you didn't show off and jump your beauty over a fence into somebody's garden, like you've done before, and then have to pay damages to keep out of jail!

MELODY

Roused by mention of his pet—disdainfully.

The damned Yankee yokels should feel flattered that she deigns to set her dainty hooves in their paltry gardens! She's a truer-born, well-bred lady than any of their women—than the one who paid us a visit this morning, for example.

SARA

Mrs. Harford was enough of a lady to put you in your place and make a fool of you.

MELODY

Seemingly unmoved by this taunt—calmly.

You are very simple-minded, my dear, to let yourself be taken in by such an obvious bit of clever acting. Naturally, the lady was a bit discomposed when she heard you and your mother coming, after she had just allowed me to kiss her. She had to pretend—

SARA

Eagerly.

She let you kiss her?

Then disgustedly.

It's a lie, but I don't doubt you've made yourself think it's the truth by now.

Angrily.

I'm going. I don't want to listen to the whiskey in you boasting of what never happened—as usual!

She puts her hands on the table and starts to rise.

MELODY

With a quick movement pins hers down with one of his.

Wait!

A look of vindictive cruelty comes into his eyes—quietly.

Why are you so jealous of the mare, I wonder? Is it because she has such slender ankles and dainty feet?

He takes his hand away and stares at her hands—with disgust, commandingly.

Keep your thick wrists and ugly, peasant paws off the table in my presence, if you please! They turn my stomach! I advise you never to let Simon get a good look at them—

SARA

Instinctively jerks her hands back under the table guiltily. She stammers.

You—you cruel devil! I knew you'd—

MELODY

For a second is ashamed and really contrite.

Forgive me, Sara. I didn't mean—the whiskey talking—as you said.

He adds in a forced tone, a trace of mockery in it.

An absurd taunt, when you really have such pretty hands and feet, my dear.

She jumps to her feet, so hurt and full of hatred her lips tremble and she cannot speak. He speaks quietly.

Are you going? I was about to tell you of the talk I had this afternoon with young Harford.

She stares at him in dismay. He goes on easily.

It was after I returned from my ride. I cantered the mare by the river and she pulled up lame. So I dismounted and led her back to the barn. No one noticed my return and when I went upstairs it occurred to me I would not find again such an opportunity to have a frank chat with Harford—free from interruptions.

He pauses, as if he expects her to be furious, but she remains tensely silent, determined not to let him know her reaction.

I did not beat about the bush. I told him he must appreciate, as a gentleman, it was my duty as your father to demand he lay his cards on the table. I said he must realize that even before you began nursing him here and going alone to his bedroom, there was a deal of gossip about your visits to his cabin, and your walks in the woods with him. I put it to him that such an intimacy could not continue without gravely compromising your reputation.

SARA

Stunned—weakly.

God forgive you! And what did he say?

MELODY

What could he say? He is a man of honor. He looked damn embarrassed and guilty for a moment, but when he found his tongue, he agreed with me most heartily. He said his mother had told him the same thing.

SARA

Oh, she did, did she? I suppose she did it to find out by watching him how far—

MELODY

Coldly.

Well, why not? Naturally, it was her duty as his mother to discover all she could about you. She is a woman of the world. She would be bound to suspect that you might be his mistress.

SARA

Tensely.

Oh, would she!

MELODY

But that's beside the point. The point is, my bashful young gentleman finally blurted out that he wanted to marry you.

SARA

Forgetting her anger—eagerly.

He told you that?

MELODY

Yes, and he said he had told his mother, and she had said all she wanted was his happiness but she felt in fairness to you and to himself—and I presume she also meant to both families concerned—he should test his love and yours by letting a decent interval of time elapse before your marriage. She mentioned a year, I believe.

SARA

Angrily.

Ah! Didn't I guess that would be her trick!

MELODY

Lifting his eyebrows—coldly.

Trick? In my opinion, the lady displayed more common sense and knowledge of the world than I thought she possessed. The reasons she gave him are sound and show a consideration for your good name which ought to inspire gratitude in you and not suspicion.

SARA

Arrah, don't tell me she's made a fool of you again! A lot of consideration she has for me!

MELODY

She pointed out to him that if you were the daughter of some family in their own little Yankee clique, there would be no question of a hasty marriage, and so he owed it to you—

SARA

I see. She's the clever one!

MELODY

Another reason was—and here your Simon stammered so embarrassedly I had trouble making him out—she warned him a sudden wedding would look damnably suspicious and start a lot of evil-minded gossip.

SARA

Tensely.

Oh, she's clever, all right! But I'll beat her.

MELODY

I told him I agreed with his mother. It is obvious that were there a sudden wedding without a suitable period of betrothal, everyone would believe—

SARA

I don't care what they believe! Tell me this! Did she get him to promise her he'd wait?

Before he can answer—bitterly.

But of course she did! She'd never have left till she got that out of him!

MELODY

Ignores this.

I told him I appreciated the honor he did me in asking for your hand, but he must understand that I could not commit myself until I had talked to his father and was assured the necessary financial arrangements could be concluded to our mutual satisfaction. There was the amount of settlement to be agreed upon, for instance.

SARA

That dream, again! God pity you!

She laughs helplessly and a bit hysterically.

And God help Simon. He must have thought you'd gone out of your mind! What did he say?

MELODY

He said nothing, naturally. He is well bred and he knows this is a matter he must leave to his father to discuss. There is also the equally important matter of how generous an allowance Henry Harford is willing to settle on his son. I did not mention this to Simon, of course, not wishing to embarrass him further with talk of money.

SARA

Thank God for that, at least!

She giggles hysterically.

MELODY

Quietly.

May I ask what you find so ridiculous in an old established custom? Simon is an elder son, the heir to his father's estate. No matter what their differences in the past may have been, now that Simon has decided to marry and settle down his father will wish to do the fair thing by him. He will realize, too, that although there is no more honorable calling than that of poet and philosopher, which his son has chosen to pursue, there is no decent living to be gained by its practice. So naturally he will settle an allowance on Simon, and I shall insist it be a generous one, befitting your position as my daughter. I will tolerate no niggardly trader's haggling on his part.

SARA

Stares at him fascinatedly, on the edge of helpless, hysterical laughter.

I suppose it would never occur to you that old Harford might not think it an honor to have his son marry your daughter.

MELODY

Calmly.

No, it would never occur to me—and if it should occur to him, I would damned soon disabuse his mind. Who is he but a money-grubbing trader? I would remind him that I was born in a castle and there was a time when I possessed wealth and position, and an estate compared to which any Yankee upstart's home in this country is but a hovel stuck in a cabbage patch. I would remind him that you, my daughter, were born in a castle!

SARA

Impulsively, with a proud toss of her head.

Well, that's no more than the truth.

Then furious with herself and him.

Och, what crazy blather!

She springs to her feet.

I've had enough of your mad dreams!

MELODY

Wait! I haven't finished yet.

He speaks quietly, but as he goes on there is an increasing vindictiveness in his tone.

There was another reason why I told young Harford I could not make a final decision. I wished time to reflect on a further aspect of this proposed marriage. Well, I have been reflecting, watching you and examining your conduct, without prejudice, trying to be fair to you and make every possible allowance—

He pauses.

Well, to be brutally frank, my dear, all I can see in you is a common, greedy, scheming, cunning peasant girl, whose only thought is money and who has shamelessly thrown herself at a young man's head because his family happens to possess a little wealth and position.

SARA

Trying to control herself.

I see your game, Father. I told you when you were drunk like this— But this time, I won't give you the satisfaction—

Then she bursts out angrily.

It's a lie! I love Simon, or I'd never—

MELODY

As if she hadn't spoken.

So, I have about made up my mind to decline for you Simon Harford's request for your hand in marriage.

SARA

Jeers angrily now.

Oh, you have, have you? As if I cared a damn what you—!

MELODY

As a gentleman, I feel I have a duty, in honor, to Simon. Such a marriage would be a tragic misalliance for him—and God knows I know the sordid tragedy of such a union.

SARA

It's Mother has had the tragedy!

MELODY

I hold young Harford in too high esteem. I cannot stand by and let him commit himself irrevocably to what could only bring him disgust and bitterness, and ruin to all his dreams.

SARA

So I'm not good enough for him, you've decided now?

MELODY

That is apparent from your every act. No one, no matter how charitably inclined, could mistake you for a lady. I have tried to make you one. It was an impossible task. God Himself cannot transform a sow's ear into a silk purse!

SARA

Furiously.

Father!

MELODY

Young Harford needs to be saved from himself. I can understand his phyical infatuation. You are pretty. So was your mother pretty once. But marriage is another matter. The man who would be the ideal husband for you, from a standpoint of conduct and character, is Mickey Maloy, my bartender, and I will be happy to give him my paternal blessing—

SARA

Let you stop now, Father!

MELODY

You and he would be congenial. You can match tongues together. He's a healthy animal. He can give you a raft of peasant brats to squeal and fight with the pigs on the mud floor of your hovel.

SARA

It's the dirty hut in which your father was born and raised you're remembering, isn't it?

MELODY

Stung to fury, glares at her with hatred. His voice quivers but is deadly quiet.

Of course, if you trick Harford into getting you with child, I could not refuse my consent.

Letting go, he bangs his fist on the table.

No, by God, even then, when I remember my own experience, I'll be damned if I could with a good conscience advise him to marry you!

SARA

Glaring back at him with hatred.

You drunken devil!

She makes a threatening move toward him, raising her hand as if she were going to slap his face—then she controls herself and speaks with quiet, biting sarcasm.

Consent or not, I want to thank you for your kind fatherly advice on how to trick Simon. I don't think I'll need it but if the worst comes to the worst I promise you I'll remember—

MELODY

Coldly, his face expressionless.

I believe I have said all I wished to say to you.

He gets up and bows stiffly.

If you will excuse me, I shall join Corporal Cregan.

He goes to the bar door. Sara turns and goes quietly out right, forgetting to clear the few remaining dishes on the center table. His back turned, he does not see her go. With his hand on the knob of the bar door, he hesitates. For a second he breaks–torturedly.

Sara!

Then quietly.

There are things I said which I regret–even now. I– I trust you will overlook– As your mother knows, it's the liquor talking, not– I must admit that, due to my celebrating the anniversary, my brain is a bit addled by whiskey–as you said.

He waits, hoping for a word of forgiveness. Finally, he glances over his shoulder. As he discovers she is not there and has not heard him, for a second he crumbles, his soldierly erectness sags and his face falls. He looks sad and hopeless and bitter and old, his eyes wandering dully. But, as in the two preceding acts, the mirror attracts him, and as he moves from the bar door to stand before it he assumes his arrogant, Byronic pose again. He repeats in each detail his pantomime before the mirror. He speaks proudly.

Myself to the bitter end! No weakening, so help me God!

There is a knock on the street door but he does not hear it. He starts his familiar incantation quotes from Byron.

"I have not loved the World, nor the World me;
I have not flattered its rank breath, nor bowed
To its idolatries a patient knee . . ."

The knock on the door is repeated more loudly. Melody starts guiltily and steps quickly away from the mirror. His embarrassment is transformed into resentful anger. He calls.

Come in, damn you! Do you expect a lackey to open the door for you?

The door opens and Nicholas Gadsby comes in. Gadsby is in his late forties, short, stout, with a big, bald head, round, florid face, and small, blue eyes. A rigidly conservative, best-family attorney, he is stiffly correct in dress and manner, dryly portentous in speech, and extremely conscious of his professional authority and dignity. Now, however, he is venturing on unfamiliar ground and is by no means as sure of himself as his manner indicates. The unexpected vision of Melody in his uniform startles him and for a second he stands, as close to gaping as he can be, impressed by Melody's handsome distinction. Melody, in his turn, is surprised. He had not thought the intruder would be a gentleman. He unbends, although his tone is still a bit curt. He bows a bit stiffly, and Gadsby finds himself returning the bow.

Your pardon, sir. When I called, I thought it was one of the damned riffraff mistaking the barroom door. Pray be seated, sir.

Gadsby comes forward and takes the chair at the head of the center table, glancing at the few dirty dishes on it with distaste. Melody says.

Your pardon again, sir. We have been feasting late, which accounts for the disarray. I will summon a servant to inquire your pleasure.

GADSBY

Beginning to recover his aplomb—shortly.

Thank you, but I want nothing, sir. I came here to seek a private interview with the proprietor of this tavern, by name, Melody.

He adds a bit hesitantly.

Are you, by any chance, he?

MELODY

Stiffens arrogantly.

I am not, sir. But if you wish to see Major Cornelius Melody, one time of His Majesty's Seventh Dragoons, who served with honor under the Duke of Wellington in Spain, I am he.

GADSBY

Dryly.

Very well, sir. Major Melody, then.

MELODY

Does not like his tone—insolently sarcastic.

And whom have I the *honor* of addressing?

As Gadsby is about to reply, Sara enters from right, having remembered the dishes. Melody ignores her as he would a servant. Gadsby examines her carefully as she gathers up the dishes. She notices him staring at her and gives him a resentful, suspicious glance. She carries the dishes out, right, to the kitchen, but a moment later she can be seen just inside the hall at right, listening. Meanwhile, as soon as he thinks she has gone, Gadsby speaks.

GADSBY

With affected casualness.

A pretty young woman. Is she your daughter, sir? I seemed to detect a resemblance—

MELODY

Angrily.

No! Do I look to you, sir, like a man who would permit his daughter to work as a waitress? Resemblance to me? You must be blind, sir.

Coldly.

I am still waiting for you to inform me who you are and why you should wish to see me.

GADSBY

Hands him a card—extremely nettled by Melody's manner—curtly.

My card, sir.

MELODY

Glances at the card.

Nicholas Gadsby.

He flips it aside disdainfully.

Attorney, eh? The devil take all your tribe, say I. I have small liking for your profession, sir, and I cannot imagine what business you can have with me. The damned thieves of the law did their worst to me many years ago in Ireland. I have little left to tempt you. So I do not see—

Suddenly an idea comes to him. He stares at Gadsby, then goes on in a more friendly tone.

That is, unless— Do you happen by any chance to represent the father of young Simon Harford?

GADSBY

Indignant at Melody's insults to his profession—with a thinly veiled sneer.

Ah, then you were expecting— That makes things easier. We need not beat about the bush. I do represent Mr. Henry Harford, sir.

MELODY

Thawing out, in his total misunderstanding of the situation.

Then accept my apologies, sir, for my animadversions against your profession. I am afraid I may be prejudiced. In the army, we used to say we suffered more casualties from your attacks at home than the French ever inflicted.

He sits down on the chair on Gadsby's left, at rear of table—remarking with careless pride.

A word of explanation as to why you find me in uniform. It is the anniversary of the battle of Talavera, sir, and—

GADSBY

Interrupts dryly.

Indeed, sir? But I must tell you my time is short. With your permission, we will proceed at once to the matter in hand.

MELODY

Controlling his angry discomfiture—coldly.

I think I can hazard a guess as to what that matter is. You have come about the settlement?

GADSBY

Misunderstanding him, replies in a tone almost openly contemptuous.

Exactly, sir. Mr. Harford was of the opinion, and I agreed with him, that a settlement would be foremost in your mind.

MELODY

Scowls at his tone but, as he completely misunderstands Gadsby's meaning, he forces himself to bow politely.

It does me honor, sir, that Mr. Harford appreciates he is dealing with a gentleman and has the breeding to know how these matters are properly arranged.

Gadsby stares at him, absolutely flabbergasted by what he considers a piece of the most shameless effrontery. Melody leans toward him confidentially.

I will be frank with you, sir. The devil of it is, this comes at a difficult time for me. Temporary, of course, but I cannot deny I am pinched at the moment—devilishly pinched. But no matter. Where my only child's happiness is at stake, I am prepared to make every possible effort. I will sign a note of hand, no matter how ruinous the interest demanded by the scoundrelly moneylenders. By the way, what amount does Mr. Harford think proper? Anything in reason—

GADSBY

Listening in utter confusion, finally gets the idea Melody is making him the butt of a joke—fuming.

I do not know what you are talking about, sir, unless you think to make a fool of me! If this is what is known as Irish wit—

MELODY

Bewildered for a second—then in a threatening tone.

Take care, sir, and watch your words or I warn you you will repent them, no matter whom you represent! No damned pettifogging dog can insult me with impunity!

As Gadsby draws back apprehensively, he adds with insulting disdain.

As for making a fool of you, sir, I would be the fool if I attempted to improve on God's handiwork!

GADSBY

Ignoring the insults, forces a placating tone.

I wish no quarrel with you, sir. I cannot for the life of me see — I fear we are dealing at cross-purposes. Will you tell me plainly what you mean by your talk of settlement?

MELODY

Obviously, I mean the settlement I am prepared to make on my daughter.

As Gadsby only looks more dumfounded, he continues sharply.

Is not your purpose in coming here to arrange, on Mr. Harford's behalf, for the marriage of his son with my daughter?

GADSBY

Marriage? Good God, no! Nothing of the kind!

MELODY

Dumfounded.

Then what have you come for?

GADSBY

Feeling he has now the upper hand—sharply.

To inform you that Mr. Henry Harford is unalterably opposed to any further relationship between his son and your daughter, whatever the nature of that relationship in the past.

MELODY

Leans forward threateningly.

By the Immortal, sir, if you dare insinuate— !

GADSBY

Draws back again, but he is no coward and is determined to carry out his instructions.

I insinuate nothing, sir. I am here on Mr. Harford's behalf, to make you an offer. That is what I thought you were expecting when you mentioned a settlement. Mr. Harford is prepared to pay you the sum of three thousand dollars—provided, mark you, that you and your daughter sign an agreement I have drawn up which specifies that you relinquish all claims, of whatever nature. And also provided you agree to leave this part of the country at once with your family. Mr. Harford suggests it would be advisable that you go West—to Ohio, say.

MELODY

So overcome by a rising tide of savage, humiliated fury, he can only stammer hoarsely.

So Henry Harford does me the honor—to suggest that, does he?

GADSBY

Watching him uneasily, attempts a reasonable, persuasive tone.

Surely you could not have spoken seriously when you talked of marriage. There is such a difference in station. The idea is preposterous. If you knew Mr. Harford, you would realize he would never countenance—

MELODY

His pent-up rage bursts out—smashing his fist on the table.

Know him? By the Immortal God, I'll know him soon! And he'll know me!

He springs to his feet.

But first, you Yankee scum, I'll deal with you!

He draws back his fist to smash Gadsby in the face, but Sara has run from the door at right and she grabs his arm. She is almost as furious as he is and there are tears of humiliated pride in her eyes.

SARA

Father! Don't! He's only a paid lackey. Where is your pride that you'd dirty your hands on the like of him?

While she is talking the door from the bar opens and Roche, O'Dowd, and Cregan crowd into the room. Mickey stands in the doorway. Nora follows Sara in from right.

ROCHE

With drunken enthusiasm.

It's a fight! For the love of God, clout the damned Yankee, Major!

MELODY

Controls himself—his voice shaking.

You are right, Sara. It would be beneath me to touch such a vile lickspittle. But he won't get off scot-free.

Sharply, a commander ordering his soldiers.

Here you, Roche and O'Dowd! Get hold of him!

They do so with enthusiasm and yank Gadsby from his chair.

GADSBY

You drunken ruffians! Take your hands off me!

MELODY

Addressing him—in his quiet, threatening tone now.

You may tell the swindling trader, Harford, who employs you that he'll hear from me!

To Roche and O'Dowd.

Throw this thing out! Kick it down to the crossroads!

ROCHE

Hurroo!

He and O'Dowd run Gadsby to the door at rear. Cregan jumps ahead, grinning, and opens the door for them.

GADSBY

Struggling futilely as they rush him through the door.

You scoundrels! Take your hands off me! Take—

Melody looks after them. The two women watch him, Nora frightened, Sara with a strange look of satisfied pride.

CREGAN

In the doorway, looking out—laughing.

Oh, it'd do your heart good, Con, to see the way they're kicking his butt down the street!

He comes in and shuts the door.

MELODY

His rage welling again, as his mind dwells on his humiliation—starting to pace up and down.

It's with his master I have to deal, and, by the Powers, I'll deal with him! You'll come with me, Jamie. I'll want you for a witness. He'll apologize to me—more than that, he'll come back here this very night and apologize publicly to my daughter, or else he meets me in the morning! By God, I'll face him at ten paces or across a handkerchief! I'll put a bullet through him, so help me, Christ!

NORA

Breaks into a dirgelike wail.

God forgive you, Con, is it a duel again—murtherin' or gettin' murthered?

MELODY

Be quiet, woman! Go back to your kitchen! Go, do you hear me!

Nora turns obediently toward the door at right, beginning to cry.

SARA

Puts an arm around her mother. She is staring at Melody apprehensively now.

There, Mother, don't worry. Father knows that's all foolishness. He's only talking. Go on now in the kitchen and sit down and rest, Mother.

Nora goes out right. Sara closes the door after her and comes back.

MELODY

Turns on her with bitter anger.

Only talking, am I? It's the first time in my life I ever heard anyone say Con Melody was a coward! It remains for my own daughter— !

SARA

Placatingly.

I didn't say that, Father. But can't you see—you're not in Ireland in the old days now. The days of duels are long past and dead, in this part of America anyway. Harford will never fight you. He—

MELODY

He won't, won't he? By God, I'll make him! I'll take a whip. I'll drag him out of his house and lash him down the street for all his neighbors to see! He'll apologize, or he'll fight, or I'll brand him a craven before the world!

SARA

Frightened now.

But you'll never be let see him! His servants will keep you out! He'll have the police arrest you, and it'll be in the papers about another drunken Mick raising a crazy row!

She appeals to Cregan.

Tell him I'm telling the truth, Jamie. You've still got some sober sense in you. Maybe he'll listen to you.

CREGAN

Glances at Melody uneasily.

Maybe Sara's right, Major.

MELODY

When I want your opinion, I'll ask for it!

Sneeringly.

Of course, if you've become such a coward you're afraid to go with me—

CREGAN

Stung.

Coward, is ut? I'll go, and be damned to you!

SARA

Jamie, you fool! Oh, it's like talking to crazy men!

She grabs her father's arm—pleadingly.

Don't do it, Father, for the love of God! Have I ever asked you anything? Well, I ask you to heed me now! I'll beg you on my knees, if you like! Isn't it me you'd fight about, and haven't I a right to decide? You punished that lawyer for the insult. You had him thrown out of here like a tramp. Isn't that your answer to old Harford that insults him? It's for him to challenge you, if he dares, isn't it? Why can't you leave it at that and wait—

MELODY

Shaking off her hand—angrily.

You talk like a scheming peasant! It's a question of my honor!

SARA

No! It's a question of my happiness, and I won't have your mad interfering— !

Desperately forcing herself to reason with him again.

Listen, Father! If you'll keep out of it, I'll show you how I'll make a fool of old Harford! Simon won't let anything his father does keep him from marrying me. His mother is the only one who might have the influence over him to come between us. She's only watching for a good excuse to turn Simon against marrying me, and if you go raising a drunken row at their house, and make a public scandal, shouting you want to murder his father, can't you see what a chance that will give her?

MELODY

Raging.

That damned, insolent Yankee bitch! She's all the more reason. Marry, did you say? You dare to think there can be any question now of your marrying the son of a man who has insulted my honor—and yours?

SARA

Defiantly.

Yes, I dare to think it! I love Simon and I'm going to marry him!

MELODY

And I say you're not! If he wasn't sick, I'd— But I'll get him out of here tomorrow! I forbid you ever to see him again! If you dare disobey me I'll— !

Beginning to lose all control of himself.

If you dare defy me—for the sake of the dirty money you think you can beg from his family, if you're his wife— !

SARA

Fiercely.

You lie!

Then with quiet intensity.

Yes. I defy you or anyone who tries to come between us!

MELODY

You'd sell your pride as my daughter— !

His face convulsed by fury.

You filthy peasant slut! You whore! I'll see you dead first— ! By the living God, I'd kill you myself!

He makes a threatening move toward her.

SARA

Shrinks back frightenedly.

Father!

Then she stands and faces him defiantly.

CREGAN

Steps between them.

Con! In the name of God!

Melody's fit of insane fury leaves him. He stands panting for breath, shuddering with the effort to regain some sort of poise. Cregan speaks, his only thought to get him away from Sara.

If we're going after old Harford, Major, we'd better go. That thief of a lawyer will warn him—

MELODY

Seizing on this—hoarsely.

Yes, let's go. Let's go, Jamie. Come along, Corporal. A stirrup cup, and we'll be off. If the mare wasn't lame, I'd ride alone—

but we can get a rig at the livery stable. Don't let me forget to stop at the barn for my whip.

By the time he finishes speaking, he has himeslf in hand again and his ungovernable fury has gone. There is a look of cool, menacing vengefulness in his face. He turns toward the bar door.

SARA

Helplessly.

Father!

Desperately, as a last, frantic threat.

You'll force me to go to Simon—and do what you said!

If he hears this, he gives no sign of it. He strides into the bar. Cregan follows him, closing the door. Sara stares before her, the look of defiant desperation hardening on her face. The street door is flung open and O'Dowd and Roche pile in, laughing uproariously.

ROCHE

Hurroo!

O'DOWD

The army is back, Major, with the foe flying in retreat.

He sees Melody is not there—to Sara.

Where's himself?

Sara appears not to see or hear him.

ROCHE

After a quick glance at her.

Lave her be. He'll be in the bar. Come on.

He goes to the bar.

O'DOWD

Following him, speaks over his shoulder to Sara.

You should have seen the Yank! His coachman had to help him in his rig at the corner—and Roche gave the coachman a clout too, for good measure!

He disappears, laughing, slamming the door behind him. Nora opens the door at right and looks in cautiously. Seeing Sara alone, she comes in.

NORA

Sara.

She comes over to her.

Sara.

She takes hold of her arm—whispers uneasily.

Where's himself?

SARA

Dully.

I couldn't stop him.

NORA

I could have told you you was wastin' breath.

With a queer pride.

The divil himself couldn't kape Con Melody from a duel!

Then mournfully.

It's like the auld times come again, and the same worry and sorrow. Even in the days before ivir I'd spoke a word to him, or done more than make him a bow when he'd ride past on his hunter, I used to lie awake and pray for him when I'd hear he was fightin' a duel in the mornin'.

She smiles a shy, gentle smile.

I was in love with him even then.

Sara starts to say something bitter but what she sees in her mother's face stops her. Nora goes on, with a feeble attempt at boastful confidence.

But I'll not worry this time, and let you not, either. There wasn't a man in Galway was his equal with a pistol, and what chance will this auld stick av a Yankee have against him?

There is a noise of boisterous farewells from the bar and the noise of an outer door shutting. Nora starts.

That's him leavin'!

Her mouth pulls down pitiably. She starts for the bar with a sob.

Ah, Con darlin', don't— !

She stops, shaking her head helplessly.

But what's the good?

She sinks on a chair with a weary sigh.

SARA

Bitterly, aloud to herself more than to her mother.

No good. Let him go his way—and I'll go mine.

Tensely.

I won't let him destroy my life with his madness, after all the plans I've made and the dreams I've dreamed. I'll show him I can play at the game of gentleman's honor too!

Nora has not listened. She is sunk in memories of old fears and her present worry about the duel. Sara hesitates—then, keeping her face turned away from her mother, touches her shoulder.

I'm going upstairs to bed, Mother.

NORA

Starts—then indignantly.

To bed, is it? You can think of sleepin' when he's—

SARA

I didn't say sleep, but I can lie down and try to rest.

Still avoiding looking at her mother.

I'm dead tired, Mother.

NORA

Tenderly solicitous now, puts an arm around her.

You must be, darlin'. It's the divil's own day for you, with all—

With sudden remorse.

God forgive me, darlin'. I was forgettin' about you and the Harford lad.

Miserably.

Oh, God help us!

Suddenly with a flash of her strange, fierce pride in the power of love.

Never mind! If there's true love between you, you'll not let a duel or anything in the world kape you from each other, whatever the cost! Don't I know!

SARA

Kisses her impulsively, then looks away again.

You're going to sit up and wait down here?

NORA

I am. I'd be destroyed with fear lying down in the dark. Here, the noise of them in the bar kapes up my spirits, in a way.

SARA

Yes, you'd better stay here. Good night, Mother.

NORA

Good night, darlin'.

Sara goes out at right, closing the door behind her.

CURTAIN

Act Four

SCENE *The same. It is around midnight. The room is in darkness except for one candle on the table, center. From the bar comes the sound of Patch Riley's pipes playing a reel and the stamp of dancing feet.*

Nora sits at the foot of the table at center. She is hunched up in an old shawl, her arms crossed over her breast, hugging herself as if she were cold. She looks on the verge of collapse from physical fatigue and hours of worry. She starts as the door from the bar is opened. It is Mickey. He closes the door behind him, shutting out an uproar of music and drunken voices. He has a decanter of whiskey and a glass in his hand. He has been drinking, but is not drunk.

NORA

Eagerly.

There's news of himself?

MALOY

Putting the decanter and glass on the table.

Sorra a bit. Don't be worryin' now. Sure, it's not so late yet.

NORA

Dully.

It's aisy for you to say—

MALOY

I came in to see how you was, and bring you a taste to put heart in you.

As she shakes her head.

Oh, I know you don't indulge, but I've known you once in a while, and you need it this night.

As she again shakes her head—with kindly bullying.

Come now, don't be stubborn. I'm the doctor and I highly recommend a drop to drive out black thoughts and rheumatism.

NORA

Well—maybe—a taste, only.

MALOY

That's the talkin'.

He pours a small drink and hands it to her.

Drink hearty, now.

NORA

Takes a sip, then puts the glass on the table and pushes it away listlessly.

I've no taste for anything. But I thank you for the thought. You're a kind lad, Mickey.

MALOY

Here's news to cheer you. The word has got round among the boys, and they've all come in to wait for Cregan and himself.

With enthusiasm.

There'll be more money taken over the bar than any night since this shebeen started!

NORA

That's good.

MALOY

If they do hate Con Melody, he's Irish, and they hate the Yanks worse. They're all hopin' he's bate the livin' lights out of Harford.

NORA

With belligerent spirit.

And so he has, I know that!

MALOY

Grins.

That's the talk. I'm glad to see you roused from your worryin'.

Turning away.

I'd better get back. I left O'Dowd to tend bar and I'll wager he has three drinks stolen already.

He hesitates.

Sara's not been down?

NORA

No.

MALOY

Resentfully.

It's a wonder she wouldn't have more thought for you than to lave you sit up alone.

NORA

Stiffens defensively.

I made her go to bed. She was droppin' with tiredness and destroyed with worry. She must have fallen asleep, like the young can. None of your talk against Sara, now!

MALOY

Starts an exasperated retort.

The divil take–

He stops and grins at her with affection.

There's no batin' you, Nora. Sure, it'd be the joy av me life to have a mother like you to fight for me–or, better still, a wife like you.

NORA

A sweet smile of pleased coquetry lights up her drawn face.

Arrah, save your blarney for the young girls!

MALOY

The divil take young girls. You're worth a hundred av thim.

NORA

With a toss of her head.

Get along with you!

Mickey grins with satisfaction at having cheered her up and goes in the bar, closing the door. As soon as he is gone, she sinks back into apprehensive brooding.

Sara appears silently in the doorway at right. She wears a faded old wrapper over her nightgown, slippers on her bare feet. Her hair is down over her shoulders, reaching to her waist. There is a change in her. All the bitterness and defiance have disappeared from her face. It looks gentle and calm and at the same time dreamily happy and exultant. She is much prettier than she has ever been before. She stands looking at her mother, and suddenly she becomes shy and uncertain—as if, now that she'd come this far, she had half a mind to retreat before her mother discovered her. But Nora senses her presence and looks up.

NORA

Dully.

Ah, it's you, darlin'!

Then gratefully.

Praise be, you've come at last! I've been sick with worry and I've got to the place where I can't bear waitin' alone, listenin' to drunks dancin' and celebratin'.

Sara comes to her. Nora breaks. Tears well from her eyes.

It's cruel, it is! There's no heart or thought for himself in divil a one av thim.

She starts to sob. Sara hugs her and kisses her cheek gently. But she doesn't speak. It is as if she were afraid her voice would give her away. Nora stops sobbing. Her mood changes to resentment and she speaks as if Sara had spoken.

Don't tell me not to worry. You're as bad as Mickey. The Yankee didn't apologize or your father'd been back here long since. It's a duel, that's certain, and he must have taken a room in the city so he'll be near the ground. I hope he'll sleep, but I'm feared he'll stay up drinkin', and at the dawn he'll have had too much to shoot his best and maybe—

Then defiantly self-reassuringly.

Arrah, I'm the fool! It's himself can keep his head clear and his eyes sharp, no matter what he's taken!

Pushing Sara away—with nervous peevishness.

Let go of me. You've hardened not to care. I'd rather stay alone.

She grabs Sara's hand.

No. Don't heed me. Sit down, darlin'.

Sara sits down on her left at rear of table. She pats her mother's hand, but remains silent, her expression dreamily happy, as if she heard Nora's words but they had no meaning for her. Nora goes on worriedly again.

But if he's staying in the city, why hasn't he sent Jamie Cregan back for his duelin' pistols? I know he'd nivir fight with any others.

Resentful now at Melody.

Or you'd think he'd send Jamie or someone back with a word for me. He knows well how tormented I'd be waiting.

Bitterly.

Arrah, don't talk like a loon! Has he ever cared for anyone except himself and his pride? Sure, he'd never stoop to think of me, the grand gentleman in his red livery av bloody England! His pride, indade! What is it but a lie? What's in his veins, God pity him, but the blood of thievin' auld Ned Melody who kept a dirty shebeen?

Then is horrified at herself as if she had blasphemed.

No! I won't say it! I've nivir! It would break his heart if he heard me! I'm the only one in the world he knows nivir sneers at his dreams!

Working herself to rebellion again.

All the same, I won't stay here the rist of the night worryin' my heart out for a man who—it isn't only fear over the duel. It's because I'm afraid it's God's punishment, all the sorrow and trouble that's come on us, and I have the black tormint in my mind that it's the fault of the mortal sin I did with him unmarried, and the promise he made me make to leave the Church that's kept me from ever confessin' to a priest.

She pauses—dully.

Go to a doctor, you say, to cure the rheumatism. Sure, what's rheumatism but a pain in your body? I could bear ten of it. It's the pain of guilt in my soul. Can a doctor's medicine cure that? No, only a priest of Almighty God—

With a roused rebellion again.

It would serve Con right if I took the chance now and broke my promise and woke up the priest to hear my confession and give me God's forgiveness that'd bring my soul peace and comfort so I wouldn't feel the three of us were damned.

Yearningly.

Oh, if I only had the courage!

She rises suddenly from her chair—with brave defiance.

I'll do it, so I will! I'm going to the priest's, Sara.

She starts for the street door—gets halfway to it and stops.

SARA

A strange, tenderly amused smile on her lips—teasingly.

Well, why don't you go, Mother?

NORA

Defiantly.

Ain't I goin'?

She takes a few more steps toward the door—stops again—she mutters beatenly.

God forgive me, I can't. What's the use pretendin'?

SARA

As before.

No use at all, Mother. I've found that out.

NORA

As if she hadn't heard, comes back slowly.

He'd feel I'd betrayed him and my word and my love for him —and for all his scorn, he knows my love is all he has in the world to comfort him.

Then spiritedly, with a proud toss of her head.

And it's my honor, too! It's not for his sake at all! Divil mend him, he always prates as if he had all the honor there is, but I've mine, too, as proud as his.

She sits down in the same chair.

SARA

Softly.

Yes, the honor of her love to a woman. I've learned about that too, Mother.

NORA

As if this were the first time she was really conscious of Sara speaking, and even now had not heard what she said—irritably.

So you've found your tongue, have you? Thank God. You're cold comfort, sitting silent like a statue, and me making talk to myself.

Regarding her as if she hadn't really seen her before —resentfully.

Musha but it's pleased and pretty you look, as if there wasn't a care in the world, while your poor father—

SARA

Dreamily amused, as if this no longer had any importance or connection with her.

I know it's no use telling you there won't be any duel, Mother, and it's crazy to give it a thought. You're living in Ireland long ago, like Father. But maybe you'll take Simon's word for it, if you won't mine. He said his father would be paralyzed with indignation just at the thought he'd ever fight a duel. It's against the law.

NORA

Scornfully.

Och, who cares for the law? He must be a coward.

She looks relieved.

Well, if the young lad said that, maybe it's true.

SARA

Of course it's true, Mother.

NORA

Your father'd be satisfied with Harford's apology and that'd end it.

SARA

Helplessly.

Oh, Mother!

Then quickly.

Yes, I'm sure it ended hours ago.

NORA

Intent on her hope.

And you think what's keeping him out is he and Jamie would take a power av drinks to celebrate.

SARA

They'd drink, that's sure, whatever happened.

She adds dreamily.

But that doesn't matter now at all.

NORA

Stares at her—wonderingly.

You've a queer way of talking, as if you'd been asleep and was still half in a dream.

SARA

In a dream right enough, Mother, and it isn't half of me that's in it but all of me, body and soul. And it's a dream that's true, and always will be to the end of life, and I'll never wake from it.

NORA

Sure, what's come over you at all?

SARA

Gets up impulsively and comes around in back of her mother's chair and slips to her knees and puts her arms about her—giving her a hug.

Joy. That's what's come over me. I'm happy, Mother. I'm happy because I know now Simon is mine, and no one can ever take him from me.

NORA

At first her only reaction is pleased satisfaction.

God be thanked! It was a great sorrow tormentin' me that the duel would come between you.

Defiantly.

Honor or not, why should the children have their lives and their love destroyed!

SARA

I was a great fool to fear his mother could turn him against me, no matter what happened.

NORA

You've had a talk with the lad?

SARA

I have. That's where I've been.

NORA

You've been in his room ever since you went up?

SARA

Almost. After I'd got upstairs it took me a while to get up my courage.

NORA

Rebukingly.

All this time—in the dead of the night!

SARA

Teasingly.

I'm his nurse, aren't I? I've a right.

NORA

That's no excuse!

SARA

Her face hardening.

Excuse? I had the best in the world. Would you have me do nothing to save my happiness and my chance in life, when I thought there was danger they'd be ruined forever? Don't you want me to have love and be happy, Mother?

NORA

Melting.

I do, darlin'. I'd give my life—

Then rebuking again.

Were you the way you are, in only a nightgown and wrapper?

SARA

Gaily.

I was—and Simon liked my costume, if you don't, although he turned red as a beet when I came in.

NORA

Small wonder he did! Shame on you!

SARA

He was trying to read a book of poetry, but he couldn't he was that worried hoping I'd come to say goodnight, and being frightened I wouldn't.

She laughs tenderly.

Oh, it was the cutest thing I've ever done, Mother, not to see him at all since his mother left. He kept waiting for me and when I didn't come, he got scared to death that his kissing me this morning had made me angry. So he was wild with joy to see me—

NORA

In your bare legs with only your nightgown and wrapper to cover your nakedness! Where's your modesty?

SARA

Gaily teasing.

I had it with me, Mother, though I'd tried hard to leave it behind. I got as red as he was.

She laughs.

Oh, Mother, it's a great joke on me. Here I'd gone to his room with my mind made up to be as bold as any street woman and tempt him because I knew his honor would make him marry me right away if—

She laughs.

And then all I could do was stand and gape at him and blush!

NORA

Oh.

Rebukingly.

I'm glad you had the dacency to blush.

SARA

It was Simon spoke first, and once he started, all he'd been holding back came out. The waiting for me, and the fear he'd had made him forget all his shyness, and he said he loved me and asked me to marry him the first day we could. Without knowing how it happened, there I was with his arms around me and mine around him and his lips on my lips and it was heaven, Mother.

NORA

Moved by the shining happiness in Sara's face.

God bless the two av you.

SARA

Then I was crying and telling him how afraid I'd been his mother hated me, Father's madness about the duel would give her a good chance to come between us; Simon said no one could ever come between us and his mother would never try to, now she knew he loved me, which was what she came over to find out. He said all she wanted was for him to be free to do as he pleased, and she only suggested he wait a year, she didn't make him promise. And Simon said I was foolish to think she would take the duel craziness serious. She'd only be amused at the joke it would be on his father, after he'd been so sure he could buy us off, if he had to call the police to save him.

NORA

Aroused at the mention of police.

Call the police, is it? The coward!

SARA

Goes on, unheedingly.

Simon was terribly angry at his father for that. And at Father too when I told how he threatened he'd kill me. But we didn't talk of it much. We had better things to discuss.

She smiles tenderly.

NORA

Belligerently.

A lot Con Melody cares for police, and him in a rage! Not the whole dirty force av thim will dare interfere with him!

SARA

Goes on as if she hadn't heard.

And then Simon told me how scared he'd been I didn't love him and wouldn't marry him. I was so beautiful, he said, and he wasn't handsome at all. So I kissed him and told him he was the handsomest in the world, and he is. And he said he wasn't worthy because he had so little to offer, and was a failure at what he'd hoped he could be, a poet. So I kissed him and told him he was too a poet, and always would be, and it was what I loved most about him.

NORA

The police! Let one av thim lay his dirty hand on Con Melody, and he'll knock him senseless with one blow.

SARA

Then Simon said how poor he was, and he'd never accept a penny from his father, even if he offered it. And I told him never mind, that if we had to live in a hut, or sleep in the grass of a field without a roof to our heads, and work our hands to the bone, or starve itself, I'd be in heaven and sing with the joy of our love!

She looks up at her mother.

And I meant it, Mother! I meant every word of it from the bottom of my heart!

NORA

Answers vaguely from her preoccupation with the police—patting Sara's hair mechanically.

Av course you did, darlin'.

SARA

But he kissed me and said it wouldn't be as bad as that, he'd been thinking and he'd had an offer from an old college friend who'd inherited a cotton mill and who wants Simon to be equal partners if he'll take complete charge of it. It's only a small mill and that's what tempts Simon. He said maybe I couldn't believe it but he knows from his experience working for his father he has the ability for trade, though he hates it, and he could easily make a living for us from this mill—just enough to be comfortable, and he'd have time over to write his book, and keep his wisdom, and never let himself become a slave to the greed for more than enough that is the curse of mankind. Then he said he was afraid maybe I'd think it was weakness in him, not wisdom, and could I be happy with enough and no more. So I kissed him and said all I wanted in life was his love, and whatever meant happiness to him would be my only ambition.

She looks up at her mother again—exultantly.

And I meant it, Mother! With all my heart and soul!

NORA

As before, patting her hair.

I know, darlin'.

SARA

Isn't that a joke on me, with all my crazy dreams of riches and a grand estate and me a haughty lady riding around in a carriage with coachman and footman!

She laughs at herself.

Wasn't I the fool to think that had any meaning at all when you're in love? You were right, Mother. I knew nothing of love, or the pride a woman can take in giving everything—the pride in her own love! I was only an ignorant, silly girl boasting, but I'm a woman now, Mother, and I know.

NORA

As before, mechanically.

I'm sure you do, darlin'.

She mutters fumingly to herself.

Let the police try it! He'll whip them back to their kennels, the dirty curs!

SARA

Lost in her happiness.

And then we put out the light and talked about how soon we'd get married, and how happy we'd be the rest of our lives together, and we'd have children—and he forgot whatever shyness was left in the dark and said he meant all the bold things he'd written in the poems I'd seen. And I confessed that I was up to every scheme to get him, because I loved him so much there wasn't anything I wouldn't do to make sure he was mine. And all the time we were kissing each other, wild with happiness. And—

She stops abruptly and looks down guiltily.

NORA

As before.

Yes, darlin', I know.

SARA

Guiltily, keeping her eyes down.

You—know, Mother?

NORA

Abruptly comes out of her preoccupation, startled and uneasy.

I know what? What are you sayin'? Look up at me!

She pulls Sara's head back so she can look down in her face—falteringly.

I can see— You let him! You wicked, sinful girl!

SARA

Defiantly and proudly.

There was no letting about it, only love making the two of us!

NORA

Helplessly resigned already but feeling it her duty to rebuke.

Ain't you ashamed to boast— ?

SARA

No! There was no shame in it!

Proudly.

Ashamed? You know I'm not! Haven't you told me of the pride in your love? Were you ashamed?

NORA

Weakly.

I was. I was dead with shame.

SARA

You were not! You were proud like me!

NORA

But it's a mortal sin. God will punish you—

SARA

Let Him! If He'd say to me, for every time you kiss Simon you'll have a thousand years in hell, I wouldn't care, I'd wear out my lips kissing him!

NORA

Frightenedly.

Whist, now! He might hear you.

SARA

Wouldn't you have said the same—?

NORA

Distractedly.

Will you stop! Don't torment me with your sinful questions! I won't answer you!

SARA

Hugging her.

All right. Forgive me, Mother.

A pause—smilingly.

It was Simon who felt guilty and repentant. If he'd had his way, he'd be out of bed now, and the two of us would be walking around in the night, trying to wake up someone who could marry us. But I was so drunk with love, I'd lost all thought or care about marriage. I'd got to the place where all you know or care is that you belong to love, and you can't call your soul your own any more, let alone your body, and you're proud you've given them to love.

She pauses—then teasing lovingly.

Sure, I've always known you're the sweetest woman in the world, Mother, but I never suspected you were a wise woman too, until I knew tonight the truth of what you said this morning, that a woman can forgive whatever the man she loves could do and still love him, because it was through him she found the love in herself; that, in one way, he doesn't count at all, because it's love, your own love, you love in him, and to keep that your pride will do anything.

She smiles with a self-mocking happiness.

It's love's slaves we are, Mother, not men's—and wouldn't it shame their boasting and vanity if we ever let them know our secret?

She laughs—then suddenly looks guilty.

But I'm talking great nonsense. I'm glad Simon can't hear me.

She pauses. Nora is worrying and hasn't listened. Sara goes on.

Yes, I can even understand now—a little anyway—how you can still love Father and be proud of it, in spite of what he is.

NORA

At the mention of Melody, comes out of her brooding.

Hush, now!

Miserably.

God help us, Sara, why doesn't he come, what's happened to him?

SARA

Gets to her feet exasperatedly.

Don't be a fool, Mother.

Bitterly.

Nothing's happened except he's made a public disgrace of himself, for Simon's mother to sneer at. If she wanted revenge on him, I'm sure she's had her fill of it. Well, I don't care. He deserves it. I warned him and I begged him, and got called a peasant slut and a whore for my pains. All I hope now is that whatever happened wakes him from his lies and mad dreams so he'll have to face the truth of himself in that mirror.

Sneeringly.

But there's devil a chance he'll ever let that happen. Instead, he'll come home as drunk as two lords, boasting of his glorious victory over old Harford, whatever the truth is!

But Nora isn't listening. She has heard the click of the latch on the street door at rear.

NORA

Excitedly.

Look, Sara!

The door is opened slowly and Jamie Cregan sticks his head in cautiously to peer around the room. His face is battered, nose red and swollen, lips cut and puffed, and one eye so blackened it is almost closed. Nora's first reaction is a cry of relief.

Praise be to the Saints, you're back, Jamie!

CREGAN

Puts a finger to his lips—cautioningly.

Whist!

NORA

Frightenedly.

Jamie! Where's himself?

CREGAN

Sharply.

Whist, I'm telling you!

In a whisper.

I've got him in a rig outside, but I had to make sure no one was here. Lock the bar door, Sara, and I'll bring him in.

She goes and turns the key in the door, her expression contemptuous. Cregan then disappears, leaving the street door half open.

NORA

Did you see Jamie's face? They've been fightin' terrible. Oh, I'm afraid, Sara.

SARA

Afraid of what? It's only what I told you to expect. A crazy row—and now he's paralyzed drunk.

Cregan appears in the doorway at rear. He is half leading, half supporting Melody. The latter moves haltingly and woodenly. But his movements do not seem those of drunkenness. It is more as if a sudden shock or stroke had shattered his coordination and left him in a stupor. His scarlet uniform is filthy and torn and pulled awry. The pallor of his face is ghastly. He has a cut over his left eye, a blue swelling on his left cheekbone, and his lips are cut and bloody. From a big raw bruise on his forehead, near the temple, trickles of dried blood run down to his jaw. Both his hands are swollen, with skinned knuckles, as are Cregan's. His eyes are empty and lifeless. He stares at his wife and daughter as if he did not recognize them.

NORA

Rushes and puts her arm around him.

Con, darlin'! Are you hurted bad?

He pushes her away without looking at her. He walks dazedly to his chair at the head of the center table. Nora follows him, breaking into lamentation.

Con, don't you know me? Oh, God help us, look at his head!

SARA

Be quiet, Mother. Do you want them in the bar to know he's come home—the way he is.

She gives her father a look of disgust.

CREGAN

'Ay, that's it, Sara. We've got to rouse him first. His pride'd nivir forgive us if we let thim see him dead bate like this.

There is a pause. They stare at him and he stares sightlessly at the table top. Nora stands close by his side, behind the table, on his right, Sara behind her on her right, Cregan at right of Sara.

SARA

He's drunk, isn't that all it is, Jamie?

CREGAN

Sharply.

He's not. He's not taken a drop since we left here. It's the clouts on the head he got, that's what ails him. A taste of whiskey would bring him back, if he'd only take it, but he won't.

SARA

Gives her father a puzzled, uneasy glance.

He won't?

NORA

Gets the decanter and a glass and hands them to Cregan.

Here. Try and make him.

CREGAN

Pours out a big drink and puts it before Melody—coaxingly.

Drink this now, Major, and you'll be right as rain!

Melody does not seem to notice. His expression remains blank and dead. Cregan scratches his head puzzledly.

He won't. That's the way he's been all the way back when I tried to persuade him.

Then irritably.

Well, if he won't, I will, be your leave. I'm needin' it bad.

He downs the whiskey, and pours out another—to Nora and Sara.

It's the divil's own rampage we've had.

SARA

Quietly contemptuous, but still with the look of puzzled uneasiness at her father.

From your looks it must have been.

CREGAN

Indignantly.

You're takin' it cool enough, and you seein' the marks av the batin' we got!

He downs his second drink—boastfully.

But if we're marked, there's others is marked worse and some av thim is police!

NORA

God be praised! The dirty cowards!

SARA

Be quiet, Mother. Tell us what happened, Jamie.

CREGAN

Faix, what didn't happen? Be the rock av Cashel, I've nivir engaged in a livelier shindy! We had no trouble findin' where Harford lived. It's a grand mansion, with a big walled garden behind it, and we wint to the front door. A flunky in livery answered wid two others behind. A big black naygur one was. That pig av a lawyer must have warned Harford to expect us. Con spoke wid the airs av a lord. "Kindly inform your master," he says, "that Major Cornelius Melody, late of His Majesty's Seventh Dragoons, respectfully requests a word with him." Well, the flunky put an insolent sneer on him. "Mr. Harford won't see you," he says. I could see Con's rage risin' but he kept polite. "Tell him," he says, "if he knows what's good for him he'll see me. For if he don't, I'll come in and see him." "Ye will, will ye?" says the flunky, "I'll have you know Mr. Harford don't allow drunken Micks to come here disturbing him. The police have been informed," he says, "and you'll be arrested if you make trouble." Then he started to shut the door. "Anyway, you've come to the wrong door," he says, "the place for the loiks av you is the servants' entrance."

NORA

Angrily.

Och, the impident divil!

SARA

In spite of herself her temper has been rising. She looks at Melody with angry scorn.

You let Harford's servants insult you!

Then quickly.

But it serves you right! I knew what would happen! I warned you!

CREGAN

Let thim be damned! Kape your mouth shut, and lave me tell it, and you'll see if we let them! When he'd said that, the flunky tried to slam the door in our faces, but Con was too quick. He pushed it back on him and lept in the hall, roarin' mad, and hit the flunky a cut with his whip across his ugly mug that set him screaming like a stuck pig!

NORA

Enthusiastically.

Good for you, Con darlin'!

SARA

Humiliatedly.

Mother! Don't!

To Melody with biting scorn.

The famous duelist–in a drunken brawl with butlers and coachmen!

But he is staring sightlessly at the table top as if he didn't see her or know her.

CREGAN

Angrily, pouring himself another drink.

Shut your mouth, Sara, and don't be trying to plague him. You're wastin' breath anyway, the way he is. He doesn't know you or hear you. And don't put on lady's airs about fighting when you're the whole cause of it.

SARA

Angrily.

It's a lie! You know I tried to stop–

CREGAN

Gulps down his drink, ignoring this, and turns to Nora–enthusiastically.

Wait till you hear, Nora!

He plunges into the midst of battle again.

The naygur hit me a clout that had my head dizzy. He'd have had me down only Con broke the butt av the whip over his black skull and knocked him to his knees. Then the third man punched Con and I gave him a kick where it'd do him least good, and he rolled on the floor, grabbin' his guts. The naygur was in again and grabbed me, but Con came at him and knocked him down. Be the mortal, we had the three av thim licked, and we'd have dragged auld Harford from his burrow and tanned his Yankee hide if the police hadn't come!

NORA

Furiously.

Arrah, the dirthy cowards! Always takin' sides with the rich Yanks against the poor Irish!

SARA

More and more humiliated and angry and torn by conflicting emotions—pleadingly.

Mother! Can't you keep still?

CREGAN

Four av thim wid clubs came behind us. They grabbed us before we knew it and dragged us into the street. Con broke away and hit the one that held him, and I gave one a knee in his belly. And then, glory be, there was a fight! Oh, it'd done your heart good to see himself! He was worth two men, lettin' out right and left, roarin' wid rage and cursin' like a trooper—

MELODY

Without looking up or any change in his dazed expression, suddenly speaks in a jeering mumble to himself.

Bravely done, Major Melody! The Commander of the Forces honors your exceptional gallantry! Like the glorious field of Talavera! Like the charge on the French square! Cursing like a drunken, foul-mouthed son of a thieving shebeen keeper who sprang from the filth of a peasant hovel, with pigs on the floor—with that pale Yankee bitch watching from a window, sneering with disgust!

NORA

Frightenedly.

God preserve us, it's crazed he is!

SARA

Stares at him startled and wondering. For a second there is angry pity in her eyes. She makes an impulsive move toward him.

Father!

Then her face hardening.

He isn't crazed, Mother. He's come to his senses for once in his life!

To Melody.

So she was sneering, was she? I don't blame her! I'm glad you've been taught a lesson!

Then vindictively.

But I've taught her one, too. She'll soon sneer from the wrong side of her mouth!

CREGAN

Angrily.

Will you shut your gab, Sara! Lave him be and don't heed him. It's the same crazy blather he's talked every once in a while since they brought him to—about the Harford woman—and speakin' av the pigs and his father one minute, and his pride and honor and his mare the next.

He takes up the story again.

Well, anyways, they was too much for us, the four av thim wid clubs. The last thing I saw before I was knocked senseless was three av thim clubbing Con. But, be the Powers, we wint down fightin' to the last for the glory av auld Ireland!

MELODY

In a jeering mutter to himself.

Like a rum-soaked trooper, brawling before a brothel on a Saturday night, puking in the gutter!

SARA

Strickenly.

Don't, Father!

CREGAN

Indignantly to Melody.

We wasn't in condition. If we had been—but they knocked us senseless and rode us to the station and locked us up. And we'd be there yet if Harford hadn't made thim turn us loose, for he's rich and has influence. Small thanks to him! He was afraid the row would get in the paper and put shame on him.

Melody laughs crazily and springs to his feet. He sways dizzily, clutching his head—then goes toward the door at left front.

NORA

Con! Where are you goin'?

She starts after him and grabs his arm. He shakes her hand off roughly as if he did not recognize her.

CREGAN

He don't know you. Don't cross him now, Nora. Sure, he's only goin' upstairs to bed.

Wheedlingly.

You know what's best for you, don't you, Major?

Melody feels his way gropingly through the door and disappears, leaving it open.

SARA

Uneasy, but consoling her mother.

Jamie's right, Mother. If he'll fall asleep, that's the best thing—

Abruptly she is terrified.

Oh God, maybe he'll take revenge on Simon—

She rushes to the door and stands listening—with relief.

No, he's gone to his room.

She comes back—a bit ashamed.

I'm a fool. He'd never harm a sick man, no matter—

She takes her mother's arm—gently.

Don't stand there, Mother. Sit down. You're tired enough—

NORA

Frightenedly.

I've never heard him talk like that in all the years—with that crazy dead look in his eyes. Oh, I'm afeered, Sara. Lave go of me. I've got to make sure he's gone to bed.

She goes quickly to the door and disappears. Sara makes a move to follow her.

CREGAN

Roughly.

Stay here, unless you're a fool, Sara. He might come to all av a sudden and give you a hell av a thrashin'. Troth, you deserve one. You're to blame for what's happened. Wasn't he fightin' to revenge the insults to you?

He sprawls on a chair at rear of the table at center.

SARA

Sitting down at rear of the small table at left front—angrily.

I'll thank you to mind your own business, Jamie Cregan. Just because you're a relation—

CREGAN

Harshly.

Och, to hell with your airs!

He pours out a drink and downs it. He is becoming drunk again.

SARA

I can revenge my own insults, and I have! I've beaten the Harfords—and he's only made a fool of himself for her to sneer at. But I've beaten her and I'll sneer last!

She pauses, a hard, triumphant smile on her lips. It fades. She gives a little bewildered laugh.

God forgive me, what a way to think of— I must be crazy, too.

CREGAN

Drunkenly.

Ah, don't be talkin'! Didn't the two of us lick them all! And Con's all right. He's all right, I'm sayin'! It's only the club on the head makes him quare a while. I've seen it often before. Ay, and felt it meself. I remember at a fair in the auld country I was clouted with the butt av a whip and I didn't remember a thing for hours, but they told me after I never stopped gabbin' but went around tellin' every stranger all my secrets.

He pauses. Sara hasn't listened. He goes on uneasily.

All the same, it's no fun listening to his mad blather about the pale bitch, as he calls her, like she was a ghost, haunting and scorning him. And his gab about his beautiful thoroughbred mare is madder still, raving what a grand, beautiful lady she is, with her slender ankles and dainty feet, sobbin' and beggin' her forgiveness and talkin' of dishonor and death—

He shrinks superstitiously—then angrily, reaching for the decanter.

Och, be damned to this night!

Before he can pour a drink, Nora comes hurrying in from the door at left front.

NORA

Breathless and frightened.

He's come down! He pushed me away like he didn't see me. He's gone out to the barn. Go after him, Jamie.

CREGAN

Drunkenly.

I won't. He's all right. Lave him alone.

SARA

Jeeringly.

Sure, he's only gone to pay a call on his sweetheart, the mare, Mother, and hasn't he slept in her stall many a time when he was dead drunk, and she never even kicked him?

NORA

Distractedly.

Will you shut up, the two av you! I heard him openin' the closet in his room where he keeps his auld set of duelin' pistols, and he was carryin' the box when he came down—

CREGAN

Scrambles hastily to his feet.

Oh, the lunatic!

NORA

He'll ride the mare back to Harford's! He'll murther someone! For the love av God, stop him, Jamie!

CREGAN

Drunkenly belligerent.

Be Christ, I'll stop him for you, Nora, pistols or no pistols!

He walks a bit unsteadily out the door at left front.

SARA

Stands tensely—bursts out with a strange triumphant pride.

Then he's not beaten!

Suddenly she is overcome by a bitter, tortured revulsion of feeling.

Merciful God, what am I thinking? As if he hadn't done enough to destroy—

Distractedly.

Oh, the mad fool! I wish he was—

From the yard, off front, there is the muffled crack of a pistol shot hardly perceptible above the noise in the barroom. But Sara and Nora both hear it and stand frozen with horror. Sara babbles hysterically.

I didn't mean it, Mother! I didn't!

NORA

Numb with fright—mumbles stupidly.

A shot!

SARA

You know I didn't mean it, Mother!

NORA

A shot! God help us, he's kilt Jamie!

SARA

Stammers.

No—not Jamie—

Wildly.

Oh, I can't bear waiting! I've got to know—

She rushes to the door at left front—then stops frightenedly.

I'm afraid to know! I'm afraid—

NORA

Mutters stupidly.

Not Jamie? Then who else?

She begins to tremble—in a horrified whisper.

Sara! You think— Oh, God have mercy!

SARA

Will you hush, Mother! I'm trying to hear—

She retreats quickly into the room and backs around the table at left front until she is beside her mother.

Someone's at the yard door. It'll be Jamie coming to tell us—

NORA

It's a lie! He'd nivir. He'd nivir!

They stand paralyzed by terror, clinging to each other, staring at the open door. There is a moment's pause in which the sound of drunken roistering in the bar seems louder. Then Melody appears in the doorway with Cregan behind him. Cregan has him by the shoulder and pushes him roughly into the room, like a bouncer handling a drunk. Cregan is shaken by the experience he has just been through and his reaction is to make him drunkenly angry at Melody. In his free hand is a dueling pistol. Melody's face is like gray wax. His body is limp, his feet drag, his eyes seem to have no sight. He appears completely possessed by a paralyzing stupor.

SARA

Impulsively.

Father! Oh, thank God!

She takes one step toward him—then her expression begins to harden.

NORA

Sobs with relief.

Oh, praise God you're alive! Sara and me was dead with fear—

She goes toward them.

Con! Con, darlin'!

CREGAN

Dumps Melody down on the nearest chair at left of the small table—roughly, his voice trembling.

Let you sit still now, Con Melody, and behave like a gintleman!

To Nora.

Here he is for ye, Nora, and you're welcome, bad luck to him!

He moves back as Nora comes and puts her arms around Melody and hugs him tenderly.

NORA

Oh, Con, Con, I was so afeered for you!

He does not seem to hear or see her, but she goes on crooning to him comfortingly as if he were a sick child.

CREGAN

He was in the stable. He'd this pistol in his hand, with the mate to it on the floor beside the mare.

He shudders, puts pistol on the table shakenly.

It's mad he's grown entirely! Let you take care av him now, his wife and daughter! I've had enough. I'm no damned keeper av lunatics!

He turns toward the barroom.

SARA

Wait, Jamie. We heard a shot. What was it?

CREGAN

Angrily.

Ask him, not me!

Then with bewildered horror.

He kilt the poor mare, the mad fool!

Sara stares at him in stunned amazement.

I found him on the floor with her head in his lap, and her dead. He was sobbing like a soul in hell—

He shudders.

Let me get away from the sight of him where there's men in their right senses laughing and singing!

He unlocks the barroom door.

And don't be afraid, Sara, that I'll tell the boys a word av this. I'll talk of our fight in the city only, because it's all I want to remember.

He jerks open the door and goes in the bar, slamming the door quickly behind him. A roar of welcome is heard as the crowd greets his arrival. Sara locks the door again. She comes back to the center table, staring at Melody, an hysterical, sneering grin making her lips quiver and twitch.

SARA

What a fool I was to be afraid! I might know you'd never do it as long as a drink of whiskey was left in the world! So it was the mare you shot?

She bursts into uncontrollable, hysterical laughter. It penetrates Melody's stupor and he stiffens rigidly on his chair, but his eyes remain fixed on the table top.

NORA

Sara! Stop! For the love av God, how can you laugh—!

SARA

I can't–help it, Mother. Didn't you hear–Jamie? It was the mare he shot!

She gives way to laughter again.

NORA

Distractedly.

Stop it, I'm sayin'!

Sara puts her hand over her mouth to shut off the sound of her laughing, but her shoulders still shake. Nora sinks on the chair at rear of the table. She mutters dazedly.

Kilt his beautiful mare? He must be mad entirely.

MELODY

Suddenly speaks, without looking up, in the broadest brogue, his voice coarse and harsh.

Lave Sara laugh. Sure, who could blame her? I'm roarin' meself inside me. It's the damnedest joke a man ivir played on himself since time began.

They stare at him. Sara's laughter stops. She is startled and repelled by his brogue. Then she stares at him suspiciously, her face hardening.

SARA

What joke? Do you think murdering the poor mare a good joke?

Melody stiffens for a second, but that is all. He doesn't look up or reply.

NORA

Frightened.

Look at the dead face on him, Sara. He's like a corpse.

She reaches out and touches one of his hands on the table top with a furtive tenderness–pleadingly.

Con, darlin'. Don't!

MELODY

Looks up at her. His expression changes so that his face loses all its remaining distinction and appears vulgar and common, with a loose, leering grin on his swollen lips.

Let you not worry, Allanah. Sure, I'm no corpse, and with a few drinks in me, I'll soon be lively enough to suit you.

NORA

Miserably confused.

Will you listen to him, Sara—puttin' on the brogue to torment us.

SARA

Growing more uneasy but sneering.

Pay no heed to him, Mother. He's play-acting to amuse himself. If he's that cruel and shameless after what he's done—

NORA

Defensively.

No, it's the blow on the head he got fightin' the police.

MELODY

Vulgarly.

The blow, me foot! That's Jamie Cregan's blather. Sure, it'd take more than a few clubs on the head to darken my wits long. Me brains, if I have any, is clear as a bell. And I'm not puttin' on brogue to tormint you, me darlint. Nor play-actin', Sara. That was the Major's game. It's quare, surely, for the two av ye to object when I talk in me natural tongue, and yours, and don't put on airs loike the late lamented auld liar and lunatic, Major Cornelius Melody, av His Majesty's Seventh Dragoons, used to do.

NORA

God save us, Sara, will you listen!

MELODY

But he's dead now, and his last bit av lyin' pride is murthered and stinkin'.

He pats Nora's hand with what seems to be genuine comforting affection.

So let you be aisy, darlint. He'll nivir again hurt you with his sneers, and his pretendin' he's a gintleman, blatherin' about pride and honor, and his boastin' av duels in the days that's gone, and his showin' off before the Yankees, and thim laughin' at him, prancing around drunk on his beautiful thoroughbred mare—

He gulps as if he were choking back a sob.

For she's dead, too, poor baste.

SARA

This is becoming unbearable for her—tensely.

Why—why did you kill her?

MELODY

Why did the Major, you mean! Be Christ, you're stupider than I thought you, if you can't see that. Wasn't she the livin' reminder, so to spake, av all his lyin' boasts and dreams? He meant to kill her first wid one pistol, and then himself wid the other. But faix, he saw the shot that killed her had finished him, too. There wasn't much pride left in the auld lunatic, anyway, and seeing her die made an end av him. So he didn't bother shooting himself, because it'd be a mad thing to waste a good bullet on a corpse!

He laughs coarsely.

SARA

Tensely.

Father! Stop it!

MELODY

Didn't I tell you there was a great joke in it? Well, that's the joke.

He begins to laugh again but he chokes on a stifled sob. Suddenly his face loses the coarse, leering, brutal expression and is full of anguished grief. He speaks without brogue, not to them but aloud to himself.

Blessed Christ, the look in her eyes by the lantern light with life ebbing out of them—wondering and sad, but still trustful, not reproaching me—with no fear in them—proud, understanding pride—loving me—she saw I was dying with her. She understood! She forgave me!

He starts to sob but wrenches himself out of it and speaks in broad, jeering brogue.

Begorra, if that wasn't the mad Major's ghost speakin'! But be damned to him, he won't haunt me long, if I know it! I intind to live at my ease from now on and not let the dead bother me, but enjoy life in my proper station as auld Nick Melody's son. I'll bury his Major's damned red livery av bloody England deep in the ground and he can haunt its grave if he likes, and boast to the lonely night av Talavera and the ladies of Spain and fightin' the French!

With a leer.

Troth, I think the boys is right when they say he stole the uniform and he nivir fought under Wellington at all. He was a terrible liar, as I remember him.

NORA

Con, darlin', don't be grievin' about the mare. Sure, you can get another. I'll manage—

SARA

Mother! hush!

To Melody, furiously.

Father, will you stop this mad game you're playing— ?

MELODY

Roughly.

Game, is it? You'll find it's no game. It was the Major played a game all his life, the crazy auld loon, and cheated only himself. But I'll be content to stay meself in the proper station I was born to, from this day on.

With a cunning leer at Sara.

And it's meself feels it me duty to give you a bit av fatherly advice, Sara darlint, while my mind is on it. I know you've great ambition, so remember it's to hell wid honor if ye want to rise in this world. Remember the blood in your veins and be your grandfather's true descendent. There was an able man for you! Be Jaysus, he nivir felt anything beneath him that could gain him something, and for lyin' tricks to swindle the bloody fools of gintry, there wasn't his match in Ireland, and he ended up wid a grand estate, and a castle, and a pile av gold in the bank.

SARA

Distractedly.

Oh, I hate you!

NORA

Sara!

MELODY

Goes on as if he hadn't heard.

I know he'd advise that to give you a first step up, darlint, you must make the young Yankee gintleman have you in his bed, and after he's had you, weep great tears and appeal to his honor to marry you and save yours. Be God, he'll nivir resist that, if I know him, for he's a young fool, full av dacency and dreams, and looney, too, wid a touch av the poet in him. Oh, it'll be aisy for you—

SARA

Goaded beyond bearing.

I'll make you stop your dirty brogue and your play-acting!

She leans toward him and speaks with taunting vindictiveness, in broad brogue herself.

Thank you kindly but I've already taken your wise advice, Father. I made him have me in his bed, while you was out drunk fightin' the police!

NORA

Frightenedly.

Sara! Hault your brazen tongue!

MELODY

His body stiffens on his chair and the coarse leer vanishes from his face. It becomes his old face. His eyes fix on her in a threatening stare. He speaks slowly, with difficulty keeping his words in brogue.

Did you now, God bless you! I might have known you'd not take any chance that the auld loon av a Major, going out to revenge an insult to you, would spoil your schemes.

He forces a horrible grin.

Be the living God, it's me should be proud this night that one av the Yankee gintry has stooped to be seduced by my slut av a daughter!

Still keeping his eyes fixed on hers, he begins to rise from his chair, his right hand groping along the table top until it clutches the dueling pistol. He aims it at Sara's heart, like an automaton, his eyes as cold, deadly, and merciless as they must have been in his duels of long ago. Sara is terrified but she stands unflinchingly.

NORA

Horror-stricken, lunges from her chair and grabs his arm.

Con! For the love av God! Would you be murthering Sara?

A dazed look comes over his face. He grows limp and sinks back on his chair and lets the pistol slide from his fingers on the table. He draws a shuddering breath—then laughs hoarsely.

MELODY

With a coarse leer.

Murtherin' Sara, is it? Are ye daft, Nora? Sure, all I want is to congratulate her!

SARA

Hopelessly.

Oh!

She sinks down on her chair at rear of the center table and covers her face with her hands.

NORA

With pitifully well-meant reassurance.

It's all right, Con. The young lad wants to marry her as soon as can be, she told me, and he did before.

MELODY

Musha, but that's kind of him! Be God, we ought to be proud av our daughter, Nora. Lave it to her to get what she wants by hook or crook. And won't we be proud watchin' her rise in the world till she's a grand lady!

NORA

Simply.

We will, surely.

SARA

Mother!

MELODY

She'll have some trouble, rootin' out his dreams. He's set in his proud, noble ways, but she'll find the right trick! I'd lay a pound, if I had one, to a shilling she'll see the day when she'll wear fine silks and drive in a carriage wid a naygur coachman behind spankin' thoroughbreds, her nose in the air; and she'll live in a Yankee mansion, as big as a castle, on a grand estate av stately woodland and soft green meadows and a lake.

With a leering chuckle.

Be the Saints, I'll start her on her way by making her a wedding present av the Major's place where he let her young gintleman build his cabin—the land the Yankees swindled him into buyin' for his American estate, the mad fool!

He glances at the dueling pistol—jeeringly.

Speakin' av the departed, may his soul roast in hell, what am I doin' wid his pistol? Be God, I don't need pistols. Me fists, or a club if it's handy, is enough. Didn't me and Jamie lick a whole regiment av police this night?

NORA

Stoutly.

You did, and if there wasn't so many av thim—

MELODY

Turns to her—grinningly.

That's the talk, darlint! Sure, there's divil a more loyal wife in the whole world—

He pauses, staring at her—then suddenly kisses her on the lips, roughly but with a strange real tenderness.

and I love you.

NORA

With amazed, unthinking joy.

Oh, Con!

MELODY

Grinning again.

I've meant to tell you often, only the Major, damn him, had me under his proud thumb.

He pulls her over and kisses her hair.

NORA

Is it kissin' my hair—!

MELODY

I am. Why wouldn't I? You have beautiful hair, God bless you! And don't remember what the Major used to tell you.

The gintleman's sneers he put on is buried with him. I'll be a real husband to you, and help ye run this shebeen, instead of being a sponge. I'll fire Mickey and tend the bar myself, like my father's son ought to.

NORA

You'll not! I'll nivir let you!

MELODY

Leering cunningly.

Well, I offered, remember. It's you refused. Sure, I'm not in love with work, I'll confess, and maybe you're right not to trust me too near the whiskey.

He licks his lips.

Be Jaysus, that reminds me. I've not had a taste for hours. I'm dyin' av thirst.

NORA

Starts to rise.

I'll get you—

MELODY

Pushes her back on her chair.

Ye'll not. I want company and singin' and dancin' and great laughter. I'll join the boys in the bar and help Cousin Jamie celebrate our wonderful shindy wid the police.

He gets up. His old soldierly bearing is gone. He slouches and his movements are shambling and clumsy, his big hairy hands dangling at his sides. In his torn, disheveled, dirt-stained uniform, he looks like a loutish, grinning clown.

NORA

You ought to go to bed, Con darlin', with your head hurted.

MELODY

Me head? Faix, it was nivir so clear while the Major lived to tormint me, makin' me tell mad lies to excuse his divilments.

He grins.

And I ain't tired a bit. I'm fresh as a man new born. So I'll say goodnight to you, darlint.

He bends and kisses her. Sara has lifted her tear-stained face from her hands and is staring at him with a strange, anguished look of desperation. He leers at her.

And you go to bed, too, Sara. Troth, you deserve a long, dreamless slape after all you've accomplished this day.

SARA

Please! Oh, Father, I can't bear— Won't you be yourself again?

MELODY

Threatening her good-humoredly.

Let you kape your mouth closed, ye slut, and not talk like you was ashamed of me, your father. I'm not the Major who was too much of a gintleman to lay hand on you. Faix, I'll give you a box on the ear that'll teach you respect, if ye kape on trying to raise the dead!

She stares at him, sick and desperate. He starts toward the bar door.

SARA

Springs to her feet.

Father! Don't go in with those drunken scum! Don't let them hear and see you! You can drink all you like here. Jamie will come and keep you company. He'll laugh and sing and help you celebrate Talavera—

MELODY

Roughly.

To hell wid Talavera!

His eyes are fastened on the mirror. He leers into it.

Be Jaysus, if it ain't the mirror the auld loon was always admirin' his mug in while he spouted Byron to pretend himself was a lord wid a touch av the poet—

He strikes a pose which is a vulgar burlesque of his old before-the-mirror one and recites in mocking brogue.

"I have not loved the World, nor the World me;
I have not flatthered uts rank breath, nor bowed
To uts idolatries a pashunt knee,
Nor coined me cheek to smiles,—nor cried aloud
In worship av an echo: in the crowd
They couldn't deem me one av such—I stood
Among thim, but not av thim . . ."

He guffaws contemptuously.

Be Christ, if he wasn't the joke av the world, the Major. He should have been a clown in a circus. God rest his soul in the flames av tormint!

Roughly.

But to hell wid the dead.

The noise in the bar rises to an uproar of laughter as if Jamie had just made some climactic point in his story. Melody looks away from the mirror to the bar door.

Be God, *I'm* alive and in the crowd they *can* deem me one av such! I'll be among thim and av thim, too—and make up for the lonely dog's life the Major led me.

He goes to the bar door.

SARA

Starts toward him—beseechingly.

Father! Don't put this final shame on yourself. You're not drunk now. There's no excuse you can give yourself. You'll be as dead to yourself after, as if you'd shot yourself along with the mare!

MELODY

Leering—with a wink at Nora.

Listen to her, Nora, reproachin' me because I'm not drunk. Troth, that's a condition soon mended.

He puts his hand on the knob of the door.

SARA

Father!

NORA

Has given way to such complete physical exhaustion, she hardly hears, much less comprehends what is said—dully.

Lave him alone, Sara. It's best.

MELODY

As another roar is heard from the bar.

I'm missin' a lot av fun. Be God, I've a bit of news to tell the boys that'll make them roar the house down. The Major's passin' to his eternal rest has set me free to jine the Democrats, and I'll vote for Andy Jackson, the friend av the common men like me, God bless him!

He grins with anticipation.

Wait till the boys hear that!

He starts to turn the knob.

SARA

Rushes to him and grabs his arm.

No! I won't let you! It's my pride, too!

She stammers.

Listen! Forgive me, Father! I know it's my fault—always sneering and insulting you—but I only meant the lies in it. The truth—Talavera—the Duke praising your bravery—an officer in his army—even the ladies in Spain—deep down that's been my pride, too—that I was your daughter. So don't— I'll do anything you ask— I'll even tell Simon—that after his father's insult to you—I'm too proud to marry a Yankee coward's son!

MELODY

Has been visibly crumbling as he listens until he appears to have no character left in which to hide and defend himself. He cries wildly and despairingly, as if he saw his last hope of escape suddenly cut off.

Sara! For the love of God, stop—let me go— !

NORA

Dully.

Lave your poor father be. It's best.

In a flash Melody recovers and is the leering peasant again.

SARA

With bitter hopelessness.

Oh, Mother! Why couldn't you be still!

MELODY

Roughly.

Why can't you, ye mean. I warned ye what ye'd get if ye kept on interferin' and tryin' to raise the dead.

He cuffs her on the side of the head. It is more of a playful push than a blow, but it knocks her off balance back to the end of the table at center.

NORA

Aroused—bewilderedly.

God forgive you, Con!

Angrily.

Don't you be hittin' Sara now. I've put up with a lot but I won't—

MELODY

With rough good nature.

Shut up, darlint. I won't have to again.

He grins leeringly at Sara.

That'll teach you, me proud Sara! I know you won't try raisin' the dead any more. And let me hear no more gab out of you about not marryin' the young lad upstairs. Be Jaysus, haven't ye any honor? Ye seduced him and ye'll make an honest gentleman av him if I have to march ye both by the scruff av the neck to the nearest church.

He chuckles—then leeringly.

And now with your permission, ladies both, I'll join me good friends in the bar.

He opens the door and passes into the bar, closing the door behind him. There is a roar of welcoming drunken shouts, pounding of glasses on bar and tables, then quiet as if he had raised a hand for silence, followed by his voice greeting them and ordering drinks, and other roars of acclaim mingled with the music of Riley's pipes. Sara remains standing by the side of the center table, her shoulders bowed, her head hanging, staring at the floor.

NORA

Overcome by physical exhaustion again, sighs.

Don't mind his giving you a slap. He's still quare in his head. But he'll sing and laugh and drink a power av whiskey and slape sound after, and tomorrow he'll be himself again—maybe.

SARA

Dully—aloud to herself rather than to her mother.

No. He'll never be. He's beaten at last and he wants to stay beaten. Well, I did my best. Though why I did, I don't know. I must have his crazy pride in me.

She lifts her head, her face hardening—bitterly.

I mean, the late Major Melody's pride. I mean, I did have it. Now it's dead—thank God—and I'll make a better wife for Simon.

There is a sudden lull in the noise from the bar, as if someone had called for silence—then Melody's voice is plainly heard in the silence as he shouts a toast: "Here's to our next President, Andy Jackson! Hurroo for Auld Hickory, God bless him!" *There is a drunken chorus of answering* "hurroos" *that shakes the walls.*

NORA

Glory be to God, cheerin' for Andy Jackson! Did you hear him, Sara?

SARA

Her face hard.

I heard someone. But it wasn't anyone I ever knew or want to know.

NORA

As if she hadn't heard.

Ah well, that's good. They won't all be hatin' him now.

She pauses—her tired, worn face becomes suddenly shy and tender.

Did you hear him tellin' me he loved me, Sara? Did you see him kiss me on the mouth—and then kiss my hair?

She gives a little, soft laugh.

Sure, he must have gone mad altogether!

SARA

Stares at her mother. Her face softens.

No, Mother, I know he meant it. He'll keep on meaning it, too, Mother. He'll be free to, now.

She smiles strangely.

Maybe I deserved the slap for interfering.

NORA

Preoccupied with her own thoughts.

And if he wants to kape on makin' game of everyone, puttin' on the brogue and actin' like one av thim in there—

She nods toward the bar.

Well, why shouldn't he if it brings him peace and company in his loneliness? God pity him, he's had to live all his life alone in the hell av pride.

Proudly.

And I'll play any game he likes and give him love in it. Haven't I always?

She smiles.

Sure, I have no pride at all—except that.

SARA

Stares at her—moved.

You're a strange, noble woman, Mother. I'll try and be like you.

She comes over and hugs her—then she smiles tenderly.

I'll wager Simon never heard the shot or anything. He was sleeping like a baby when I left him. A cannon wouldn't wake him.

In the bar, Riley starts playing a reel on his pipes and there is the stamp of dancing feet. For a moment Sara's face becomes hard and bitter again. She tries to be mocking.

Faith, Patch Riley don't know it but he's playing a requiem for the dead.

Her voice trembles.

May the hero of Talavera rest in peace!

She breaks down and sobs, hiding her face on her mother's shoulder—bewilderedly.

But why should I cry, Mother? Why do I mourn for him?

NORA

At once forgetting her own exhaustion, is all tender, loving help and comfort.

Don't, darlin', don't. You're destroyed with tiredness, that's all. Come on to bed, now, and I'll help you undress and tuck you in.

Trying to rouse her—in a teasing tone.

Shame on you to cry when you have love. What would the young lad think of you?

CURTAIN